Praise for The Inventor's Bible

"I've been an inventor for over 30 years and I believe this book could have saved me much time and heartbreak had I read it when I was beginning my invention career. Please read it and take notes! It is a God send!"

—Stanley I. Mason, inventor of granola bars and disposable diapers, and holder of 55 patents

"This remarkable and useful book is also a pleasure to read. David Sarnoff defined an inventor as someone who makes others wealthy. Inventors who read *The Inventor's Bible* and follow its step-by-step instructions can now more easily reap the rewards of their own ingenuity. Mr. Docie has covered virtually every facet of invention: securing proprietary rights, patents, licensing, and production. A must for inventors."

—Hugh Downs, network news anchor

"Inventors should add *The Inventor's Bible* to their bookshelves immediately, but be certain to keep it within easy reach. This book is a powerful compilation of useful advice from one who's already been down the rocky road to invention commercialization. Ron Docie has a lot of experience to share, and he does so in an organized fashion."

—Don Kelly, former director, U.S. Patent and Trademark Office

"*The Inventor's Bible* is an excellent collection of practical information useful to all inventors, based on the author's personal, extensive experience. The text is not presented from a legal viewpoint but rather follows essentially a business and financial path, which an inventor who hopes to be financially successful should consider."

—Don Banner, former commissioner of Patents and Trademarks

"Docie presents a refreshingly frank approach to invention development, protection, and marketing. His advice on 'tiered risk' alone may save inventors—both neophytes and seasoned—thousands of dollars as well as invaluable time. His book is a must for the desk of every serious inventor."

—Jack Lander, VP, Yankee Invention Exposition and former president, United Inventors Association

"Rarely does a book come along that provides an accurate and thorough explanation of the invention process. The real life stories and quotes not only provide for interesting reading, but also adequately warn inventors of potential pitfalls commonly encountered. I unconditionally recommend *The Inventor's Bible* to all inventors and businesses."

—MICHAEL S. NEUSTEL, PATENT ATTORNEY AND DIRECTOR, NATIONAL INVENTOR FRAUD CENTER

"I can recommend *The Inventor's Bible* without reservation as a most valuable resource for the independent inventor who is looking to develop and bring an invention to market. *The Inventor's Bible* clearly reflects in-depth and long-time experience in the field of invention development and marketing. It is well written, very readable, and quite thorough in describing what needs to be done by the independent inventor ... to successfully get an invention into the marketplace. The 'how-to' aspects of *The Inventor's Bible* are extraordinarily detailed."

—GEORGE LEWETT, FORMER DIRECTOR OF TECHNOLOGY EVALUATION AND ASSESSMENT, U.S. DEPARTMENT OF COMMERCE

"*The Inventor's Bible* takes readers from Genesis through Revelation in the convoluted world of inventing. This is the book that every inventor needs to read. Ron's been there and done that, and now he shares his expertise. Before spending a dime on patents or services, get this book."

—ED SOBEY, PH.D., PRESIDENT, NORTHWEST INVENTION CENTER

"Ron has outdone himself! In my opinion *The Inventor's Bible* can be considered *the* good book for inventors. What you have here is an excellent guide with rock-solid information that will definitely steer inventors on the right course. But it's also intertwined with Ron's personal experiences, which will help inventors be more realistic. Whether you're thinking of inventing or are already a professional in the inventing field, Ron's book is easy to read and he covers a lot of territory very well. Chock full of information, tips, words of wisdom, and resources, *The Inventor's Bible* will teach, prod, and motivate readers."

—STEPHEN PAUL GNASS, FOUNDER, INVENTION CONVENTION® TRADE SHOW AND PRESIDENT, NATIONAL CONGRESS OF INVENTOR ORGANIZATIONS

THE INVENTOR'S BIBLE

THE INVENTOR'S BIBLE

4th EDITION

HOW TO MARKET AND LICENSE YOUR BRILLIANT IDEAS

RONALD LOUIS DOCIE, SR.

TEN SPEED PRESS
Berkeley

DISCLAIMER

The author is not an attorney, nor registered to practice in the U.S. Patent and Trademark Office, and it is not expressed or implied that the author or Docie Development, LLC, is offering legal advice in this book. The author, acting in association with Docie Development, LLC, dba Docie Marketing, refers to legal issues, the law, patenting, and U.S. Patent and Trademark Office procedures only as a point of reference to help readers better understand how legal issues may affect the marketing aspect of invention commercialization. The laws and procedures referred to in this book may change, and various courts may interpret the law differently. Therefore, the author highly recommends that the reader should seek the assistance of registered attorneys regarding legal issues, the law, and patent procedures.

Neither the author nor the publisher endorse or condone the use of any of the resources listed in this book—these references are merely provided as potential sources of useful information and material. The author is neither responsible for the content of these external resources nor for the consequences that may result from readers' interactions therewith. It is the reader's responsibility to research and confirm the legitimacy and relevance of any resources he or she decides to contact. Also note that some of the resources' contact details and information may have changed since the time of publication.

Published in the United States by Ten Speed Press, an imprint of the Crown Publishing Group,
a division of Penguin Random House LLC, New York.
www.crownpublishing.com
www.tenspeed.com

Ten Speed Press and the Ten Speed Press colophon are registered trademarks of Penguin Random House LLC.

Previous editions of this work were published by Ten Speed Press, Berkeley, CA, in 2001, 2004, and 2010.

Special thanks to Jim Davie of the Patent Office Historical Collection for granting permission to use the original patent drawings (appearing on pages 1, 11, 31, 51, 83, 113, 145, 170, 175, 185, 195, 197, 223, 228, and 232) from the National Archives.

Portions of this book previously appeared in *Royalties in Your Future: An Inventor's and Entrepreneur's Guide to Marketing and Licensing Inventions, Patents, and Technology*, copyright © 1998 by Ronald L. Docie Sr., Docie Marketing.

A downloadable PDF of the Patent and New Product Marketing Workbook (see page 267) is available at www.penguinrandomhouse.com. Simply look up *The Inventor's Bible* to access the computer printable version.

Library of Congress Cataloging-in-Publication Data
Docie, Ronald Louis.
The inventor's bible : how to market and license your brilliant ideas / Ronald Louis Docie, Sr.—Fourth edition.
pages cm
Summary: "The definitive guide for inventors, newly updated with the latest patenting laws, information on crowdfunding, and online resources"—Provided by publisher.
Includes bibliographical references and index.
1. New products—United States—Marketing. 2. Inventions—United States. 3. Patents—United States. I. Title.
HF5415.153.D63 2015
658.8--dc23
2015014587

Trade Paperback ISBN: 978-1-60774-927-1
eBook ISBN: 978-1-60774-928-8

Printed in the United States of America

Cover design by Katy Brown
Text design by Lynn Bell, Monroe Street Studios
Workbook design by Betsy Stromberg

10 9 8 7 6 5 4 3 2 1

Fourth Edition

Dedication

I dedicate this book to Drue, Taylor, Ronnie, Tanner, Louis, and Andy Docie, Metra, and, of course, Mom. Your love and support helped to make this possible. Thank you!

Acknowledgments

To the people listed below, please know that your assistance, support, and in many cases kindness, over the years are much appreciated. I thank each of you for your help: Aaron Wehner; Al and Dorothy Shuster; Ashley Thompson; Basia Lubicz; Becky Huff; Betsy Stromberg; Bob Fletcher, Esq.; Brad Liston; Brady Kahn; Cal Wight; Carol Oldenburg; Cathie Kirik; Chris Eaton; Cindy Poland; Clancy Drake; Clark Yamazaki; Colleen Cain; Connie Elliot; Craig Dupler; Dave Ellies; David Brennan; David Smith; Dick Apley; Don Armacost, Jr.; Don Banner, Esq.; Don Barber; Don Wirtshafter; Donald Grant Kelly; Douglas Duff, Esq.; Ed Verry; Ed Zimmer; Eric Hart; Erika Bradfield; Ernie Clutter; George P. Lewett; Greg Ball; Jack Lander; Jeffery Jones; Jennifer McArdle; Jennifer Mortiz; Jerry Udell; Jim Davie; Jim Fingar; Jo Ann Deck; Joanne Hayes-Rines; Joe Marsalka; John Brody, Esq.; John Buck Patton; John L. Gray, Esq.; John Larimer; John Schenken; Jonathan Gilbert; Julie Bennett; Kara Van de Water; Kathy Adkins; Kathy Harty; Kathy Hashimoto; Katy Brown; Kirsty Melville; Lisa Regul; Lisa Y. Dill; Lajos Silberstein; Lorena Jones; Lynn Bell; Marilyn Rauch; Melody Sands Gates; Michael Neustel, Esq.; Mike Bayless; The Misers; Norm and Abe Matthew; Paul Gerig, Esq.; Patrick O'Reilly III, Esq.; Peter Charet; Peter Trzyna; Phil Wood; Pixie Picketts; Richard Maulsby; Richard Shoemaker; Rick Leveille; Robert D. Smith; Robert Wortman; Ron Versic; Sam Girton; Sam Seltzer; Sara Madsen; Scott Pennelly; Shirley Coe; Stanley I. Mason; Susan Pi; T. Lee Van Dyke; Ted Ongaro Family; Tommy Tabatowski; Todd H. Bailey, Esq.; Wally Downs; and all of my clients, past and present, from around the world.

And thanks to those organizations whose help I greatly appreciate: Frognet; Keglar, Brown, Hill & Ritter; Publisher's Graphics; U.S. Department of Energy; USPTO Office for Independent Inventors; Inventor's Initiative Program for financial support for the original work that provided the basis for this book; the USDOE Inventions and Innovation program for being there for inventors; the U.S. Patent and Trademark Office for permissions; and all the people behind the scenes at Ten Speed Press who helped to make this book possible.

Contents

Foreword

When inventors dream up their first invention, they are usually very secretive about it for fear that someone else will steal it, patent it, and become wealthy. After the third or fourth invention, they sadly realize the problem is not that someone will steal the invention, but that no one really cares about it. The world has gotten along quite well without their inventions, and they're met with the attitude of "who needs it?"

The introduction of virtually every invention has been met with this same attitude—from the electric light bulb to inside plumbing. More often than not, inventors and discoverers die before their works are recognized and put to use. It has been said that if a man builds a better mousetrap the world will beat a path to his door. Unfortunately, it may be the door to his mausoleum rather than his laboratory

In *The Experts Speak* by Christopher Cerf and Victor Navasky, it is interesting to find that most important inventions were at first ridiculed by contemporary scientists. Nearly all of the things we have become accustomed to were summarily put down by the very cream of the scientific community—luminaries such as H. G. Wells, William Pickering, Lord Kelvin, Thomas Edison, and Antoine Lavoisier were willing to dismiss as useless or unworkable such things as electric lights, rockets, airplanes, telephones, and radio transmission. We should consider this when we blindly accept the word of experts who have no better track record in prediction than you or me.

Possibly the most astonishing negative prediction was made after a demonstration of the phonograph. The phonograph would be of no commercial value, revealed Thomas Edison, the phonograph's inventor. The only use Edison could think of for the phonograph was the recording of wills, completely overlooking the potential for the multibillion-dollar entertainment industry. (Ironically, there is no evidence that the phonograph was ever used to record wills.)

David Sarnoff once defined an inventor as someone who makes others wealthy. This may be a bit cynical, but more often than not, it is also true. Fortunately, inventors and discoverers are driven by motives other than purely financial ones, although monetary reward would be welcomed by all of them.

In addition to demystifying the complex process of commercializing an invention, *The Inventor's Bible* will help channel a greater portion of the earned money to the inventor, where it belongs. With the help of experienced individuals like Ronald Docie, the path to success, while not guaranteed, is clearer than it's ever been. In the end we are all inventors to a greater or lesser extent. Our dreams will become the next great inventions, those inventions will be refined again and again, and we'll continue to create wondrous things we don't even know to wish for yet.

—J. W. Downs

June, 2001

INTRODUCTION

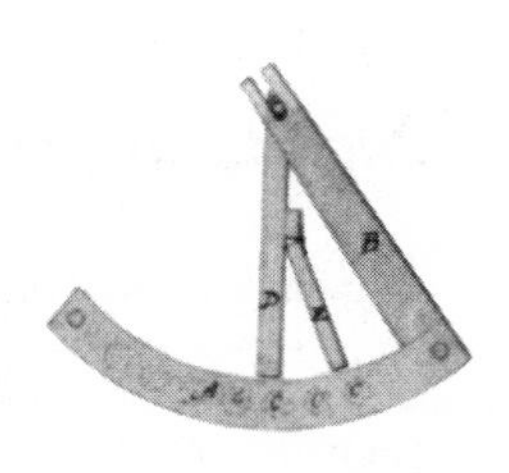

IT'S NOT THE WILL TO WIN,
BUT THE WILL TO PREPARE TO WIN
THAT MAKES THE DIFFERENCE.

—Bear Bryant

When it comes right down to it, profiting from inventions can be quite simple. All you have to do is determine who wants your invention and find out what companies will develop it into a product, approach these companies and establish a mutually satisfactory value and compensation basis for your invention, and finally sip margaritas on the tropical island of your choice. Okay, that last bit is probably somewhat unrealistic, though there are a few who have accomplished such feats.

Really, though, the process of commercializing your invention and receiving royalties does not have to be complicated. Mostly, it involves good, old-fashioned common sense; a realistic, methodical approach; the ability to communicate effectively with others; and plenty of hard work and perseverance.

This book will help you focus your common sense and develop a realistic, workable plan for commercializing your invention. It will show you how effective communication with a network of industry contacts will help you research your market, target potential business partners, and strike a good deal for your inventions. You'll have to supply the hard work and perseverance, but as an inventor, you already know all about those.

Washing machine by B. Hinckley (U.S. Patent No. 6357X), 1831.

Ultimately society benefits from good inventions. Though new inventions are not necessary for existence, some inventions make life on the planet better for people and for the environment. The planet's population is not getting any smaller, and population growth alone will create new challenges and problems in the years to come, necessitating new solutions. This gives inventors a sort of open season for the foreseeable future.

Inventions can only provide a benefit if they come into commercial use. *The Inventor's Bible* will help you convey your valuable knowledge and developments to others so society can benefit and you can gain fair and just remuneration for your ideas.

The Climate for Independent Inventors

Forty years ago, many corporations had substantial research and development budgets, and the NIH (Not Invented Here) syndrome was prevalent throughout the country, for that matter throughout the world. The NIH syndrome is characterized by the arrogant belief that no one can improve on the company's own research and development efforts; therefore, companies turned away outside inventors. If it was "not invented here," they didn't want it.

In the 1970s corporations became very competitive, and budgets for research and development were among the first to be slashed. As a result, in the 1980s corporations were starving for new products and technologies. There were corporate buyouts; when a company was losing ground in its market, it often bought out a division of another company that had compensatory sales velocity. Corporations also started to show some interest in inventions from outside sources.

In the 1990s and now, invention licensing is at an all-time high. Licensing is in vogue. Corporations have departments for licensing in and licensing out. Many large corporations are now offering disclosure agreements and welcoming submissions from outside inventors.

One of the biggest turn-offs for companies is being approached by uninformed inventors with unrealistic expectations. Inventors often submit inventions without doing their homework; they are notorious for submitting inventions to the wrong type of companies. This wastes everyone's time. I will tell you how to identify and communicate with companies in the appropriate industries to help you ascertain: whether there is a market for your invention, what the perceived value is for your invention in the marketplace, which of those companies may be appropriate to commercialize your invention, and how to structure the best deal to maximize your potential profits.

Why Did I Write This Book?

There is a great deal of information available on the subject of patenting and negotiations. However, detailed information about how to get from product development to finding manufacturers and licensees is largely missing. There is also a general lack of information for inventors on some of their most vital concerns: What is my invention worth? What steps should I take first? Is free government help available? Who can I trust, and how can I keep from getting ripped off?

When I invented my own safety product for automobiles at the ripe age of twenty-one, I had all the same concerns. I proceeded down an arduous path, asking others, "What do you do when you think up an invention?" Within three years, I was capitalized, and the product was on the market. My success brought me into contact with other inventors who sought assistance, so I began Docie Marketing, an organization dedicated to helping independent inventors make it down the rocky road of invention development. I now have twenty-five years of experience as a successful inventor and invention development consultant and can provide a first-hand account of the licensing process to even greater numbers of inventors with this book than I can in my business.

What Is in This Book?

The Inventor's Bible is a primer for beginners and a detailed overview for experienced inventors and entrepreneurs. It is an in-depth how-to manual on the commercialization process: how to research the market for your invention, how to find manufacturers and potential licensees, how to develop a licensing and commercialization strategy, how to identify risks, how to effect commercialization on a low budget, and how to select professionals to help you.

This book explains how to generate money from your invention through licensing. If you want to start a business or commercialize your invention on your own, this book will show you how to develop a realistic market projection, learn the competitive conditions in your industry, identify market and financial risks, and assess other factors important to an inventor or entrepreneur.

As I know from experience, it can be hard for inventors to make the leap from drafting table to marketplace. Chapter 1 is a reality check to help you start looking at your invention in the light of its marketability and its licensability—two things that unfortunately have little to do with whether an invention works or is a fresh idea. This chapter also considers the pros and cons of starting your own invention-based business versus licensing your invention to a manufacturer.

Chapter 2 deals with patent strategy, challenging the notion that applying for a patent is always the first thing an inventor must do before starting the marketing process. Chapter 3 cracks the commercialization code, showing you how to find people in your trade who can provide you with help in the commercialization process. The techniques in this chapter and the next three are at the heart of what I do as an invention development professional.

Chapters 4, 5, and 6 teach you how to target companies that can make and distribute your invention. As you zero in on these companies, these chapters will guide you through protecting your rights, understanding a company's perspectives, and getting the best possible deal or deals for your invention. Chapter 7 advises you on finding professionals who can help you manage the process.

Throughout this book, numerous sidebars highlight tips and tidbits of information that will help round out your perspective on this process, and provide insider strategies and techniques that may come in handy. I offer insight gleaned from my years in the profession in sidebars marked with this symbol . Sidebars that contain a strategy or technique have this icon next to them . Three case studies are threaded throughout the book, with a segment following the Introduction and each chapter. These real-life examples provide interesting stories of life in the commercialization trenches and teach lessons about both successes and failures.

The end of the book is chock full of information and resources. The appendices include a list of invention evaluation criteria, a sample confidential disclosure agreement, a risk/reward ratio test, a quick-reference flow chart of the invention commercialization process, an inventor's questionnaire, information about patents and patenting, and a helpful glossary of terms. The extensive inventor-oriented resources section lists free government programs, sources of grant money, useful websites, comprehensive databases, inventor's organizations, relevant publications, conferences, and much more.

The Meaning of Success

What determines whether an invention will be a success or failure? Achieving success is like climbing a ladder. One step is finding and contracting with the manufacturers that will produce your invention. The next successful level is to have your invention distributed to the marketplace. Another step may be to actually receive royalties for your invention. Yet another step may be to receive more money for your invention than what you paid out. Ultimately, inventors would like to see their invention put in the hands of all those people who could use it. I hope *The Inventor's Bible* helps you climb to the top of your ladder.

ANALOGOUS TERMS

Many terms are used synonymously in this book. For the subtle nuances of these terms, please refer to the Glossary.

The following terms broadly refer to your invention:

- intellectual property
- proprietary property
- proprietary technology
- proprietary rights
- intellectual rights
- innovation
- prototype
- new product
- trade secret
- know-how
- technology
- working model

The rights to your invention may be in the form of a/an:

- patent
  - design patent
  - utility patent
  - letters patent
- trademark
  - trade dress
- trade secret
- copyright

The process of attempting to get a patent issued:

- applying for a patent
- patent pending
- patent application
- patent prosecution
- prosecuting a patent
- provisional application

A lawyer is also known as an/a:

- attorney
- patent attorney
- legal counsel
- counsel
- patent practioner

To exploit your invention in business, you may:

- commercialize
- market
- license
- sell
- assign
- transfer technology
- monetize

A knowledgeable person with experience in your industry is a/an:

- company executive
- industry executive
- key decision maker
- key player
- key person
- key executive

Your industry or product field is the:

- industry
- market
- marketplace
- trade
- field

And finally, manufacturers and licensees are sometimes generically referred to as *companies*.

CASE STUDY

An Automotive Accessory (Part 1)

The first invention I ever commercialized was my own. It was a blind spot, stick-on mirror for vehicles. Unlike round fish-eye mirrors that have been available for exterior rearview mirrors since 1963, my invention incorporates a wedge-angle into the base of the mirror housing. The illustration on this page is from a patent drawing.

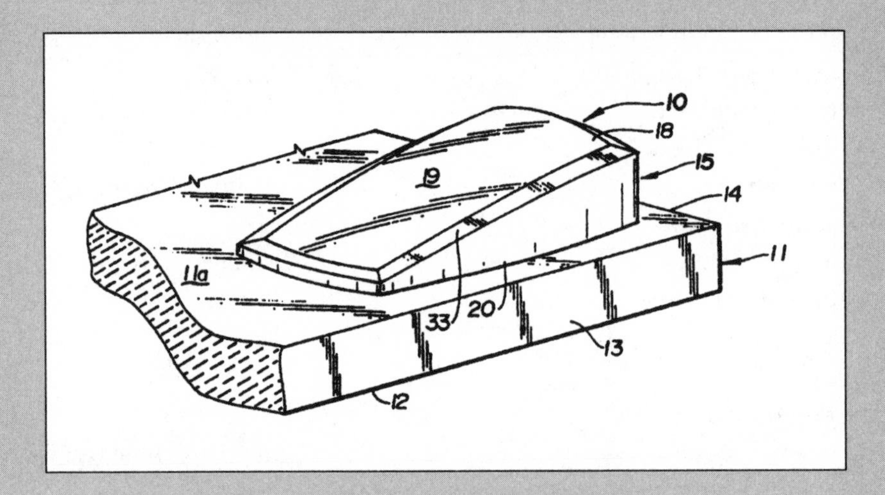

The wedge-angle in my invention reduces the blind spot by 50 percent. Because of its rectangular design, my Docie wedge mirror takes up little space on your existing rearview mirror. Your existing rearview mirror is more functional than with other stick-on mirrors on the market.

The Docie wedge mirror was conceived in the fall of 1976 when I was driving a hearse in Columbus, Ohio. A car snuck into my blind spot when I was about to change lanes. I had to swerve to miss it and, after doing so, looked into my mirror and thought, "There ought to be an extra piece of glass wedged into the corner of this mirror so you can see a car in that blind spot." My next thought was, "Why don't I invent a mirror for that?"

I had no experience with inventions, so I asked a co-worker and a relative what I should do. The relative said to get a patent, and the co-worker suggested that I talk with the president of an innovative manufacturing company he knew.

I considered both pieces of advice, but first talked with a local patent attorney. He said that it would be hard to get a patent because other mirrors on the market were too similar to my invention. If an invention is too similar to other things already in existence, your patent application will be rejected because it is considered "obvious" to design such a small improvement.

The patent attorney did inform me of the price to apply for a patent and said he would do that for me. He suggested that before talking with anyone else about the exact nature or secrets of my invention, I should have them sign a confidential disclosure agreement. He gave me a sample to use so I could customize one for myself.

With the disclosure agreement in hand, I proceeded to the president of the innovative manufacturing company. He was inventor friendly and had licensed other outside inventions before. He signed my disclosure agreement. We talked about the trials and tribulations of developing inventions, and at the end of the conversation, he signed a receipt that indicated the exact nature of the invention we talked about. He referred me to the president of an auto parts company, who went through the same procedure with me and in turn referred me to a designer who did the same thing. The designer ultimately produced drawings of the invention for me at no charge. These proved quite useful in helping to illustrate my invention and in getting quotes on the cost of production.

In four or five months, I had about a dozen signed disclosure agreements and had obtained enough information to be totally confused about what to do next. I could not afford the total cost of patenting, nor could I afford the high cost of tooling to produce the invention myself. Questions arose such as, "Should I raise money to try to manufacture it myself?" or "Should I find a company that will foot the expenses and pay me a royalty?" or "Is there some in-between ground or combination of the above?" I did approach some blind-spot mirror manufacturers; however, none of them expressed any interest in putting up enough money to get the project started.

I decided to publicly disclose the Docie wedge mirror at an inventor's workshop and show and spent a few hundred dollars filing a patent application. With the help of an inventor-friendly lens manufacturer, I developed my first decent prototype. The show went well, I thought. My invention received an enthusiastic response, and I met Alex, a successful entrepreneur and speaker at one of the workshops who became my business partner.

Alex and I contemplated whether it would be better to manufacture and market my invention ourselves or to license it to a company that would pay royalties. We also considered importing the product from Taiwan and reselling it in the United States. Although there was no guarantee that a patent would ever be issued, we proceeded under the assumption that it would be.

Regardless of which route we decided to take, some money would be necessary for developing, testing, and test marketing. Alex agreed to pay up to $25,000 for the development of both the product and test marketing; in return I offered him one-half undivided interest in any rights, such as patent rights and trademarks, that might result from my invention.

It became apparent that no manufacturer wanted to dive into this project at the time. The cost of tooling would be high, and although the invention was an improvement over existing products, it would not net much more money because it was replacing an existing product. In fact, my invention would have to be priced higher. Several manufacturers noted that there was no demand for the product. This is understandable since with a new invention, no one knows to ask for it. Most of these companies produce hundreds, if not thousands, of different types of automotive accessories. To consider a major investment in a product that would sell for $1 to $2, a company wants significant assurances that there will be sales.

Having my invention manufactured in Taiwan and shipped to the United States would be cheaper, but it would take at least six months to get the initial units into the country. Just to get samples for test marketing, we would have to incur the cost of tooling the molds for the mirror's housing and its convex surface. In Taiwan the molds were much cheaper than here, but we would have to order several thousand units. If we wanted to change the design after our initial test market, we would be stuck with thousands of units that were no good.

With all of these dilemmas, we decided that we needed to test the market to determine consumer acceptance for the product and have temporary molds made to get an accurate market indication from potential consumers. We ended up paying somewhere in the neighborhood of $5,000 for our initial test molds.

Our test marketing at a local department store went well, and we were encouraged to design packaging and proceed with permanent molds. We procured orders for a few thousand mirrors from local auto parts stores and regional warehouses. We even swung a manufacturing and distribution deal with a Canadian automotive accessory manufacturer that covered the cost of our permanent molds and allowed us to offer the product in the United States at a reduced cost because of currency differences in our favor.

With polished, finished products, we obtained national distribution in Canada, where the competition was less and the mirror's improvements were embraced with greater enthusiasm. This positive reception eventually spread to the United States as well.

Subsequently, we negotiated a deal with Rubber Queen, a major U.S. manufacturer and marketer of auto accessories, and before we knew it the molds were being moved from Toronto to Jackson, Ohio. Rubber Queen did a great job of producing promotional material, a full-color package, and a full range of four different mirror sizes to fit small cars, large cars, pickup trucks, vans, and large trucks. The shape and proportion of the mirrors remained the same; only the size changed. We had extensive sales of the Docie wedge mirror in Canada, Great Britain, and, to a lesser extent, in the United States. Next, the one-two punch.

This is the end of the first segment of the wedge mirror story. As you will read in the ensuing chapters, this invention project took some unpredictable twists.

To be continued in chapter 2 . . .

HINDSIGHT LESSONS

My first big hurdle was my own lack of knowledge and experience. Had I not been willing to learn from and accept the advice of others who had much more experience than I did, this project would never have been successful. I utilized the assistance of professionals, starting with patent attorneys. I learned how to communicate with others without revealing trade secrets.

Another hurdle was to promote a product that no one seemed to want. Let's face it, if your invention is truly unique and novel, people don't know that they want it because they don't even know it exists. Therefore, the manufacturers to which you want to sell your invention won't have seen any market demand for your invention.

If you cannot find a company that is willing to take a risk on your invention, your only alternative may be to develop it yourself. Many companies are started to promote an individual invention. I did this until I was finally able to license my product to larger companies after proving that there was a market demand for it. If you have the talent and resources, you can do this on your own; otherwise, you will need to enlist the help of an entrepreneur.

CHAPTER ONE

COMMERCIALIZATION 101

Moving from the Drafting Table to the Store Shelf

IF YOU TOOK AWAY EVERYTHING IN
THE WORLD THAT HAD TO BE INVENTED,
THERE'D BE NOTHING LEFT EXCEPT A LOT
OF PEOPLE GETTING RAINED ON.

—Tom Stoppard

This chapter contains some basic truths about commercializing your invention. It can take some serious mental adjustment—and a whole new set of skills—to get an invention from the drawing board to the marketplace. Right now, no one values your invention as much as you do. And you may learn that your estimation of its value is unrealistic. Your first and most important task in the commercialization process is to discover your invention's true value and its perceived value in the marketplace (these may be two different things). Let me explain.

The Dilemma of Perceived Value

Value is in the eye of the beholder. When inventors ask me whether I think their invention has value, I can unequivocally answer yes, because at the very least the invention has value to one person—the inventor. The question isn't whether an invention has value. The questions are: how many other people perceive the value of the invention, who are they, where are they, who could supply it to them, and how much will they pay for it?

Plough by Stephen McCormick (U.S. Patent No. 4325X), Fauquier County, Virginia, 1826.

A good example of the dilemma of perceived value is the seat belt. When seat belts were introduced to the marketplace, they failed miserably. Few people voluntarily purchased them. In fact, there is an adage in the automotive industry, which holds true in other industries, that "safety does not sell." Seat belts did not come of age until the federal government mandated their installation. I seriously doubt that the inventor ever received a royalty, considering the time it took to get acceptance in the market. The inventor saw the true value of the seat belt when he invented it; however, the general public did not perceive that value at that time.

Some of the things that could benefit humankind the most are not accepted even when they are free. For example, it doesn't cost much to exercise more, watch our diet, and take steps to reduce stress. We know these actions could help us extend our life span and improve our quality of life, yet how many of us are as prudent as we could be about implementing them? Bear this in mind when you have developed an invention that you know will help people. The customer may not perceive any value in your invention. Most inventors face this problem of perceived value.

Be realistic about the demand for your invention.

YOU CAN'T CONTROL MARKET DEMAND

When you market your invention to the masses, you are at the mercy of their whims. Even large companies such as Proctor and Gamble and General Electric, with their massive advertising budgets, cannot ram products down consumers' throats. When General Electric originally introduced its line of energy-efficient appliances years ago, they met with failure. The appliances benefitted the environment and provided real value to consumers; they could pay for themselves in savings. The consumers, however, were not willing to pay 10 percent more up front to receive a 50 percent savings over a three-year period.

HABITS ARE HARD TO CHANGE

Most inventions today are nonessential. The vast majority of new inventions do not address people's basic needs for food, shelter, or clothing. Most are improvements to technology that we lived without seventy years ago. Not only is your invention probably nonessential, it may require users or manufacturers and distributors to change their habits.

Most inventions also necessitate a change, and people tend to resist change—especially nonessential change. The marketing staff is going to have to peddle another product, the retailers are going to have to put another product on the shelf, and the consumer is going to be faced with yet another buying decision.

To convince all of these people to go to such a bother, an invention must show that it leads to significant enough savings or a significant enough improvement to warrant the change. Sometimes a 10 percent improvement is not enough.

Sometimes a 30 percent improvement is required to change the modus operandi. Your invention may save 10 percent of the energy used in the United States, which can considerably reduce our dependence on a nonrenewable energy source. However, individual consumers may not be willing to change their way of doing things unless it offers at least a 20 percent improvement.

Is the Clock Ticking?

Inventors tend to work in a vacuum. It is not uncommon for inventors to come to me with patents issued five or ten years ago. My first question is, "What have you been doing for the last decade?"

When it comes to profiting from your intellectual property, the clock is ticking. The more you delay, the greater the chance that someone else will develop a similar or superior invention. There are also time constraints in the patenting process. Timing is a very important aspect of invention development. Being first and fastest doesn't always work; many inventors have developed revolutionary concepts only to die paupers—and then their invention makes it into the mainstream years later. However, for most inventors, a sense of some urgency is important. You should keep your project moving forward as best you can (until your research suggests otherwise), either by following the steps in this book or by hiring someone to do it for you.

Choose the path that requires the least resources and will take the least amount of time.

Prototype, Patent, or Market: Which to Do First?

A dilemma facing many inventors is what to do first: make a prototype, apply for a patent, or determine the marketability of their inventions. Generally, the best answer is to do whichever one requires the least resources and can be done in the least amount of time.

For example, if you have a very complicated and capital-intensive invention, such as a hybrid engine that requires expensive, exotic alloys, you may be faced with a cost of $200,000 just to produce the prototype. The patent may cost you $3,000 initially, and the time and effort to do the market research could cost $2,000. You can make your decision based on a *tiered risk:* spend no more time and money than it takes to determine that your invention is not worth pursuing further. Most inventions never become commercially viable, for any number of reasons beyond the inventor's control. If you determine your market feasibility first and find that there is little chance of your invention being accepted in the marketplace, then don't go to the expense of patenting or prototyping.

Or perhaps your invention is very simple and can be prototyped with off-the-shelf material found in a hardware store and produced in your garage within a week. Your initial industry research (which I'll talk about later in the book) may indicate that the market is highly dependent on consumer preference and dependent on packaging, so your product may require some test marketing to see whether it will work at the retail level. Producing the product and having it packaged and test marketed may be a fairly expensive proposition, requiring a budget of $30,000 or more to do it properly. In this case, you might make a prototype and do a patent search. After the patent search submit a patent application and then either proceed with test marketing yourself or try to interest potential licensees in bearing the costs of test marketing.

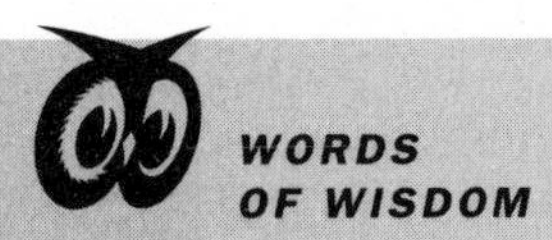

WORDS OF WISDOM

You may find that you must make a substantial investment and actually manufacture an invention to prove its commercial viability and to interest potential licensees. In this case, keep careful track of your expenses and constantly weigh these expenses against any potential royalty. The higher your investment, the more important good market research is. You may even decide it's worth your while to hire marketing and investment development professionals. You may eventually use your expenses as deductions against income from your invention—so save your receipts. (More about that in Chapter 7.)

Research Your Industry: The Real First Step

Developing a prototype, applying for a patent, or doing market research does not always cost a lot of time and money. Each case is unique. Eventually you will need to learn more about your industry and the factors that will affect the commercialization of your invention. Get to know your industry by exploring its distribution channels and interviewing key people at every level.

HELP IS YOURS FOR THE ASKING

As long as you rely on others to help you commercialize your inventions, you are at the mercy of those factors that influence them. This reality is probably the hardest lesson for inventors to appreciate. But working with others is not all bad—not by a long shot. Perceived value, market demand, and consumer habits are somewhat predictable. Manufacturers have direct experience in dealing with these challenging factors. You don't have to reinvent the wheel in your quest to commercialize your invention. Your industry is full of knowledgeable people who can help you—often for free.

I firmly believe that if the hundreds of inventors who have approached me over the past quarter century had simply broadcast to the world that they had an invention and asked whatever company wanted the product to voluntarily pay them a fair compensation, then as a whole, that group of inventors would be further ahead and richer. What usually happens is that inventors are overly paranoid, worrying about sharing information and dwelling on areas of development that are inappropriate. Consequently, their inventions never see the light of day.

Don't get me wrong—I believe it is extremely important for inventors to protect their intellectual property. There is a law of diminishing returns, however. At some point you need to trust others, particularly experts in the field of invention commercialization, to help you. The trick is to share information on a need-to-know basis and with no more people than necessary to get the job done. *The Inventor's Bible* will show you how to be selective and share information prudently.

Interviewing key industry members uncovers valuable information. Through the interview process, I discovered over time that three out of four of my clients' inventions were not worth pursuing further. I once represented a person who invested over $50,000 to patent and start limited production of his invention. He then hired me to find a licensee at a national trade show. Within two hours at the trade show, after talking with just five key members of the industry, I learned that there was an obscure federal law that prohibited the use of his invention. I later verified this through legal counsel. Obviously it would have paid this inventor to have done some industry research before spending $50,000! If these laws were expected to change during the life of a patent, however, it may have been worthwhile to pursue the patent rights. This was another factor to be weighed in this situation.

KEY DECISION MAKERS

Industry experts can help you with your projects. One of the least understood and most underutilized resources is people in your industry or trade. If you talk to the right person, you can learn vital information about the market, manufacturing processes and costs, market risks, and sales potential; determine the value of your invention; and obtain referrals to other key people who can tell you more. One of your contacts may even become your champion—someone within a company who is looking out for your best interests and can help guide your invention to a successful commercialization. I call these people "key decision makers."

EVALUATION CRITERIA

Inventors generally aren't aware of all the factors that go into commercializing an invention. Jerry Udell, Ph.D., of Southwest Missouri University, developed thirty-three criteria for evaluating the commercialization potential of inventions. These criteria have been broadly accepted as state of the art for evaluating inventions. The most recent version of these criteria (there are forty-one now) is found in Appendix C. The list is exhaustive and informative—and can be a little sobering.

When key decision makers in companies provide you with a critical evaluation of your invention, the response will lean in one of three different ways. Either it is favorable and they want to proceed with you, it is not favorable and they do not recommend proceeding, or the basis of your invention is sound but it needs more market research or specific changes to make it acceptable. If they have a keen interest in your innovation, they may be willing to proceed with the prototyping, patenting, and market feasibility assessment at no cost to you.

Who are these key people, and how does one go about finding them? As a rule, they are people with years of experience in their field who hold positions of substantial authority. They include presidents and owners of companies; vice presidents of marketing, sales, or engineering; general managers; and in the case of distributors and retailers, managers and buyers. They have closely followed trends and activities in their trade, attend their national and regional trade shows, and can anticipate changing market conditions. Their advice can be invaluable. The following chapters outline steps you can take to make all the right connections in your industry and utilize these key people effectively.

Learn to trust others—prudently.

Is Licensing Your Invention the Way to Go?

Selling an invention is similar to selling a house. When you sell your house, you transfer your title, and someone else is in charge of and liable for the house from that point on. When you sell your invention, the scenario is the same except it is called *assigning* rather than selling. You, the inventor, would be the *assignor*, and the person receiving the title or ownership of your patent would be the *assignee*.

Instead of selling, you may choose to *lease* your house. In this case you retain the title to your house and give someone permission to use it for a limited period of time. In consideration for this, they will pay you on a regular basis. The terms of this lease are entirely up to you and the person leasing your house to negotiate within the boundaries of the law. With inventions and other intellectual property, we use the term *license* instead of lease.

You have likely heard the term license used in various contexts. The most common license, a driver's license, gives you permission to drive a car on public streets in exchange for agreeing to certain terms and conditions. Another form of licensing is practiced in the promotional and merchandising industry. Here, images of Big Bird may be placed on tennis shoes, and Big Bird gets a worm for each tennis shoe sold. This, of course, is contingent on Big Bird's good behavior and positive public exposure.

Licensing an invention is similar in practice. You are offering a manufacturer, for example, the right to manufacture and sell your invention for a period of time, and in consideration for this they will pay you on a quarterly basis. You are the *licensor*, and the manufacturer is the *licensee*. It is up to the parties to negotiate the terms of the license within the boundaries of antitrust laws and other regulations that would affect licenses and similar business arrangements.

SHOULD I SELL OR LICENSE?

You will generally have a better chance of licensing your invention than assigning (selling) your rights for two reasons. First, it is hard to ascertain what the eventual value of an invention will be. This almost invariably results in a win/lose situation. If the value is estimated high, the inventor wins and the assignee company loses. On the other hand, if the estimates are low, the inventor loses out.

Second, companies do not like to pay cash up front unless they absolutely have to or unless they are flush with cash and want to control the entire development process. Generally when a company makes a commitment to manufacture and promote an invention, it is already anticipating a substantial financial commitment for tooling, manufacturing setup, engineering expense, advance purchase of raw materials, marketing, and promotional expense. Paying you cash in addition to these expenses increases the burden on the company and depletes resources it could otherwise use to promote your invention. Licensing is a way for the company to conserve cash flow and expenses.

Licensing has its own advantages. When you assign (sell) your invention, you normally lose control of it. Although you may have cash in hand from the sale of your invention, the company has the prerogative to ditch your technology and simply sit on it unless you've made other arrangements, such as a clause in the contract that states it reverts to you if the company doesn't produce it. It may be as important to you to see your invention commercialized as it is to receive cash from it. Having an invention commercialized can give you a substantial head start in attracting interest in additional inventions. This may eventually be worth more than the initial cash you would have received from your first commercialized invention.

The key advantage of assigning your intellectual property rights (patent) to someone else is that you get cash in hand sooner than if you go the licensing path. Some inventors, who believe their invention is "the next best thing" and worth $10 million, will turn down a smaller offer to hold out for more money. Because of the risks inherent in licensing any invention, even "the next best thing," the inventor may ultimately end up with nothing. As you will read in the upcoming chapters, patents are vulnerable and the risks of commercialization are high, even under the best conditions. So if a company offers you a good amount of cash, think twice before thumbing your nose at it.

The guy who invented the first wheel was an idiot. The guy who invented the other three, he was a genius.

—SID CAESAR

SHOULD I GO IT ALONE?

Now let's look at the pros and cons of licensing versus starting your own company.

There are numerous variables at play during the commercialization of a technology, including the company, the management, the technology, and the marketing team. The more variables you introduce, the greater your risk of failure. If you start a new company with a new product, your chance of success is much less than with an existing company that is already established in the field with experience and knowledge in a similar product line. For example, 3M is an experienced company noted for its marketing savvy and its ability to commercialize new inventions. Yet the rumor is that seven out of ten new 3M products fail. With all its resources, its success rate is only 30 percent.

Because of the startup risks, it is important to seriously investigate the distinct advantages of having your invention introduced by an existing company with experience in your field. The company can effectively promote your invention. A skilled sales force can go straight to the clients you hope will buy your invention. This can greatly reduce the amount of time it takes to introduce your invention to the marketplace. What you lose in control when you license can be gained tenfold from a timing standpoint.

Licensing provides another strong advantage when it is time to sell your manufactured invention to customers. Manufacturers that introduce a single invention or a small product line often have a very hard time selling to large accounts. Large retail outlets prefer to deal with companies where they can do one-stop shopping. Buyers (or purchasing agents) for the big outlets want to reduce the number of bills they get and the number of vendors they see each week. It is particularly challenging for a startup company to introduce a new invention.

Few products have an unlimited life cycle. In time, your invention may be replaced by new technology. What will your company sell then? Most single-item companies that are still around after five years have introduced new products and expanded their product line. Companies need new products to survive. Several clients have asked me for help in licensing their invention after they have already sold a few thousand units. After a year or two, they had discovered the headaches associated with starting a one-item company. They came to recognize the efficiencies of having their invention championed by a larger, established company.

Don't dive in unless you know the water's depth.

On the other hand, starting your own company may be the only way to go—if only to eventually attract a licensee. If you have attempted the assignment or licensing route and no manufacturer is interested in your invention at its current stage of development, you may need to do a small test market (which I'll discuss in Chapter 3) and a limited production run to prove your invention has sales potential.

Even if you choose to manufacture your invention yourself, the information found in the following chapters will greatly enhance your efforts. Collecting the market research you need to identify potential licensees provides the same information you need to write the marketing section of your business plan. As you assemble your research, you may find yourself on the cutting edge of your industry. You may ultimately have newer information than some of the key industry players you befriend. When you have up-to-date information to bring to the table, you will hold an executive's interest, which is important to getting an entrepreneurial effort started.

Marketability versus Licensability

Marketability refers to how well your invention is received by the market (or consumers), regardless of who sells it. Your invention may have a high potential for gross sales. However, if you don't have a good patent position, you may not make any profit from it, or another company could design around you.

Licensability refers to what your intellectual property rights are worth to a manufacturer (or licensee). You may have a solid patent and get a great licensing deal, but this doesn't guarantee that consumers will buy it.

To profit from licensing your invention, it must be *both* marketable and licensable. A substantial focus of this book is on helping you determine both the marketability and licensability, and to avoid confusing the two.

IS IT MARKETABLE?

The first basic question to answer is whether your invention/product is marketable. Here are some of the things I look for in a marketable invention:

- How is your invention better than what's on the market? Is it less expensive; does it have more options; is it more efficient, longer lasting, easier to use? And how much so?
- How crowded is the market for items like yours? Is your product sufficiently different to warrant taking up shelf space in a store?
- How easy is it to accept your product? Will it require a radical change or a long learning curve? Does it align with accepted engineering principles? Is it in a product area that typically accepts changes or improvements?
- Is the market large enough to support the amount of anticipated investment for development and promotion of your product?
- Is market timing sensitive? Is this the right window of opportunity? Will your product's life cycle (its longevity in the market) be adequate to pique the interest of distributors? In other words, is it a short fad or a longer trend?
- Can your product be reasonably produced in an acceptable price range?
- Are there other known attempts to market similar items that have met with success or failure?

IS IT LICENSABLE?

What makes an invention licensable? Your intellectual property position can be a major factor in this regard. Even though your invention may be very marketable, manufacturers will be less likely to license your invention when their competitors can easily circumvent your patent. Here are some of the top factors that affect licensability:

- What is the scope of proprietary protection for your product, and what is its relative importance in this product area?
- Which companies control this market, and what is their predisposition to embracing outside inventions?
- Are there companies in this market that are capable of doing the necessary development for your product? Do you have the resources to fill in the gaps?
- Are you willing to work with others and to accept the reality of your invention's market potential and its true worth to the industry?
- Will your invention generate enough revenue to support the efforts of the licensee? Is there room for profit for everyone along the channels of distribution?
- Have any negative precedents been set with similar products that might cause companies to shy away from inventions like yours?

As you can see, a product's marketability may have very little to do with its licensability. In other words, a company may successfully market and profit from an invention, with or without a patent, and with or without competition. It is common for companies to have successful product lines without any particular patent position. In these cases, the company's market position and its established distribution pave the way for success. Such a scenario minimizes the worth of your patent position and renders your invention less licensable.

But just because your invention is not very licensable does not mean that you will not potentially profit from it. You may still find a willing licensee for it. Or you may start your own company, have it manufactured, or blend your product into an existing company, if you have one. There are hundreds of books and resources for entrepreneurs; join a local entrepreneur's club, contact the U.S. Small Business Administration, browse libraries and bookstores, or check the resources section of this book for more information. You may also have other intellectual property, such as know-how or trade secrets, which you can license or sell. We'll talk about those in the next section.

What Do You Have That You Can License?

This seems like a pretty simple question, but it reveals one of the least understood aspects of licensing: you may have other types of proprietary protection besides a patent. You might have a trademark, know-how, trade secrets, or a copyright. Let's look at one invention and see how each of these proprietary rights can be incorporated into a licensing package.

Say you have invented a printing press. With the *apparatus* itself, you may have patentable subject matter that covers the rollers and inkers and other parts and how they fit into the machine. This would fall under the claims of your *utility patent.* You may also claim the method by which the rollers and inkers work together to lay down newsprint. This claim for how the apparatus works would fall within the method claims of your utility patent.

If your printing press has a unique design feature, you may be able to get a *design patent,* which would strictly cover the ornamental and visual design, regardless of how the printing press works.

Know the different types of protection available.

You may have gained a particular knowledge about the printing process that leads to more efficient use of your printing press. This would be part of the *know-how* you bring to someone who wants to use your technology. This know-how generally is not patentable; however, it is still a valuable asset.

Now let's say that you have developed a formula for ink to be used with your printing press. It is a secret formula that could not be reverse engineered. This would be a *trade secret,* a valuable commercial asset. You can license trade secrets just as you can license your patent rights.

You have also developed a name for your printing press, the PRINT-ALL. You have registered a logo with the U.S. Patent and Trademark Office (USPTO) after performing a trademark search to make sure that no one else in your field uses it. Your *registered trademark* is also a valuable commercial asset that can be included in your package of proprietary rights.

You have also created instructions with illustrations showing the operational features and proper use of your printing press. These instructions are your copyrightable material. Although you automatically became owner of the *copyright* when you produced your work, registering your copyright with the U.S. Library of Congress may permit you to seek greater damages and attorney fees in the event you pursue an infringer. Your copyright will last through your lifetime and seventy years beyond.

Last but not least, you are also making your time available for *consulting* to anyone interested in licensing your technology. Once they license your technology, you will be available to work for an hourly fee to answer questions pertaining to your technology. Consulting is another valuable asset that can be included in your licensing package of goodies.

It will be important to a potential licensee to know about all your proprietary items included in your total licensing package. You may enter into a licensing agreement with a company that does not have the engineering and development resources for your project, and you may fill in this gap. In this case, your total package can be as important as specific patent rights.

PATENTING BUSINESS METHODS

Patenting "business methods" is a misnomer because one can patent only processes, machines, manufactures, or compositions of matter that are new, useful, and not obvious. Thus, it is better to think of patents as indirectly protecting business innovations; for example, a computer programmed to carry out certain operations, a method of using such a computer, software on a disk, and so on.

General business concepts or models that portray a vision or strategy or merely human activity for a business are not patentable. Instead, consider the method claims outlined in the printing press example.

High-profile court cases that have brought the notion of patenting "business methods" came to the forefront with Merrill Lynch's CMA account, Amazon.com's patent on one-click shopping, and Priceline.com's patent on reverse auctions.

While there has been a significant increase in the number of patents in this field, this is a specialized area of patent practice in computer program–related inventions where knowledge in finance, insurance, tax, real estate, and e-commerce can be invaluable.

Whom Can You License To?

The chart in Chapter 3 shows a product moving from its original point of manufacturing through the various distribution steps to the consumer. The product may be designed and engineered by a U.S. manufacturer that produces the item in the United States in its own facility or subcontracts to a submanufacturer. Or the item may be produced in another country and imported. In any of these scenarios you, as an independent inventor seeking a licensee, will be dealing with a U.S. manufacturer. This is the entity with which you will initiate a license contract. There are several reasons for this.

First, manufacturers, whether they produce in the United States or import, may provide engineering support to work with both you and company personnel responsible for production and quality control. Inventions often require engineering modifications to increase manufacturing efficiency.

A manufacturer is also responsible for capital expenses, including any tooling, quality control, coordination of production schedules with delivery of raw materials, and service. Additionally, the manufacturer is usually responsible for packaging. Manufacturing and packaging are often an integrated process. Even when products are acquired from overseas, there is generally a U.S. company that acts as a middleman and calls itself a manufacturer.

Second, major retailers and wholesalers require the manufacturers that supply them to carry product liability insurance. Since manufacturers are usually responsible for quality control, they are in the best position to assure that the basic design and performance specifications are maintained and to catch any defects.

Since licensing almost always takes place at the manufacturer level, manufacturers, not distributors, are more likely to be set up to pay royalties. A progressive manufacturer will see an outside inventor's contribution as a complement to the company's research and development staff. Other types of licensing variations are discussed in later chapters.

So What Does It Take?

Several elements must work together for licensing to be a viable alternative. Even if you end up starting your own company, the following factors will likely be required for profitability:

- A willing market. The question isn't whether an invention has value or not; the question is how many people will perceive that value and how many of those will be willing to pay the cost. Markets can be very tricky, and they don't always make sense (or cents), but you should try to understand them.
- A manufacturer willing and capable of manufacturing your invention and reimbursing you with royalties, a cash buyout, a consulting fee, or other fair remuneration.
- Appropriate intellectual property position: i.e., patents, trade secrets, know-how, formulas, or unique designs.

Getting patent protection, identifying a market, and finding a manufacturer may require other things: you may need to test market your invention or develop a working prototype, find a patent attorney or get help from an invention marketing company, explore several industries, or brush up on your negotiating skills.

Invention consists in avoiding the constructing of useless contraptions and in constructing the useful combinations which are in infinite minority. To invent is to discern, to choose.

—JULES HENRI POINCARÉ

YOUR COMMITMENT

How seriously do you take this invention project? How much effort and resources are you willing to expend? Where there's a will, there's usually a way, but will the end gain make your efforts worth it?

If the majority of the inventors I've met relentlessly pursued the commercialization of their inventions, most would have likely achieved some kind of market presence. The question you need to ask yourself is whether or not the resulting income would be worth the effort. All inventors think their products are going to be a great success, but there are lots of chances for failure along the way.

The following chapters will take you through the process of evaluating the worth of your invention, weighing the risks involved in commercialization, and deciding when you should move forward and when you should throw in the towel.

Keep It in Perspective

Perhaps 2 percent of all patents are commercialized, and fewer than this make enough money to pay for the initial expenses. Most successful patents are owned by major corporations, which often patent something for formality's sake that they fully intend to market regardless. This fact stacks the odds even further against the independent inventor. Think positive, but don't overestimate the ultimate value of your invention. There are too many sad stories of inventors pouring money into inventions that never provide a return on their investment.

Know when to let go of a particular invention.

You take a risk whenever you spend time and money on your invention. Remember to approach the commercialization process with an effort to minimize your risks. This will make it easier to bail out if and when that becomes necessary.

Dropping a project that you've been working on for years can be very painful, but at some point it may be the smart thing to do. Think of it this way: You are an idea person, and you'll come up with another invention that might be successful. Remembering this and keeping the commercialization process in perspective will save you time, money, and the personal energy you'll need for future successes.

SIMPLIFIED STEPS

For a quick reference guide to the invention commercialization process through licensing, refer to the Simplified Steps chart in Appendix E.

THINGS TO REMEMBER AND CONSIDER

- Perceived value. What do others think your invention is worth?
- Time is of the essence; be aware of deadlines.
- To do or not to do: Patent? Prototype? Marketing?
- To do or not to do: Sell? License? Manufacture yourself?
- What factors affect your invention's marketability and licensability?
- What intellectual property do you have to offer?
- Identify your strengths and weaknesses.

CASE STUDY
Tire Technology

In 1982 I met with an investor in Columbus, Ohio, who was concerned because he had invested over $600,000 in a project and hadn't received even a nibble on the technology, let alone made any money. The inventor's son had also invested $600,000 in a project. Ten years and $1.2 million had gone by, and the inventor had only a very nice prototype to show for it. I was called in to see whether anything could be salvaged.

The man's invention was a state-of-the-art stationary road wheel that could measure levels of vibrations or harmonics in a vehicle tire. The technology allowed car manufacturers to learn more about the dynamics of tires. The inventor had received patents on this technology and on a supporting technology. Between the two technologies, he could measure inconsistent high mass areas in a tire, grind off the mass to make the tire more uniform, and reduce rolling resistance by 6 percent—all while the tire was moving down the production line. If a manufacturer could reduce tires' rolling resistance by 6 percent, it would save the company hundreds of millions of dollars in CAFE requirement fines. (CAFE standards are federal mandates for fuel efficiency. If a line of cars doesn't meet the standards, the car manufacturer is fined for that model that year.)

Needless to say, this is not the type of technology you would find in your local department store. I immediately asked the inventor, "Has anyone ever expressed an interest in this technology?" and "Who?" The inventor said AS Co. once considered the technology. He "knew" they weren't interested because they never got back to him. This is fatal mistake No. 1. Some inventors think manufacturers will stop everything to work with them. This couldn't be further from the truth.

During my research of the industry, I learned that only two companies in the world produced tire uniformity road wheels for tire manufacturers' production lines. These two were AS Co. and a German company. As it turned out, AS Co. commanded an 80 percent market share in the United States and the majority of the worldwide marketshare. In other words, for every ten machines in the United States, eight are from AS Co. This company also happened to be located within a three-hour drive of us.

I called the vice president at AS Co., whom the inventor had dealt with in the past. He had a very positive memory of the inventor's project and wondered what the current status was. He had been waiting to hear from the inventor because the last time they spoke the inventor needed to do a bit more development to polish the invention. Neither the inventor nor the company executive did any type of follow-up, each thinking the other should be doing it. I can't tell you how many times I have seen this happen.

I told the vice president that the inventor's technology was ready and the inventor believed he could reduce tires' rolling resistance by 6 percent. The company executive was very excited, although guardedly so. Being an engineer himself, he needed to see raw testing results. He agreed to visit Columbus to see the inventor's technology in person. The meeting went very well, and we agreed to take the steps necessary to proceed with AS Co.

AS Co. was keenly interested in this technology partly because its sales were stagnant. It had already saturated the market with its tire uniformity machines, and the only market left was replacing and servicing its machines. The company was looking for avenues to increase sales in an apparently dying business. It was considering two different projects, the inventor's tire uniformity retrofit and a retrofit computerization of the existing tire uniformity machines currently in operation in the field. The market for retrofitting with the inventor's technology would potentially be $600 million. Needless to say, there was a lot at stake here.

The major automobile manufacturers, and primarily General Motors, set the standards for tire manufacturers. To make this technology fly, GM had to be convinced that it would help them. We arranged a meeting with a manager at the GM proving grounds upon AS Co.'s referral. The manager said GM wanted to see results from an independent source, a test set up by an experienced engineer or consultant who would organize an impartial and adequate test.

We hired an independent engineering consultant from Detroit who specialized in tires and computer programs and arranged a test at UniRoyal's facilities near Toledo, Ohio. To proceed with the testing, however, the inventor had to pay additional expenses, and he was at the end of his rope financially. AS Co. was not ready to commit to licensing the technology until it knew GM was on board.

We needed a cash infusion—and when such a dilemma arises it's best to look to those entities that will ultimately benefit if things go well. In this case, AS Co. was the appropriate candidate. Since they were not in a position to initiate a license, we approached them regarding a license option. AS Co. agreed to pay $25,000 for an option, or first right, to exercise a reasonable license agreement between the parties in the event the test results were positive. The results were to be double-checked and verified at GM's proving grounds in Michigan.

The test was inconclusive. Although there was an improvement in the rolling resistance of some tires, the same results were not achieved in every instance, and it could not be concluded to what degree the inventor's technology affected the improvement. Nor could the test predict what degree of change the inventor's technology would make in the rolling resistance of tires in years to come. After all, manufacturers are constantly improving their technology for manufacturing tires. The net result was an average 2 to 3 percent improvement, which moved AS Co.'s perception of the invention from exceptional to marginal.

We attended meetings for the Society of Automobile Engineers (SAE) to learn the latest developments in tire technology and to generate interest from engineers in the tire industry. After all, these were the people who controlled the specifications for tire production. We found momentum going toward research and development to improve the tire manufacturing process, rather than fix the tires after they come out of the mold. AS Co. also sought the input from tire industry engineers and received the same response. AS Co. chose not to exercise its option to license the inventor's technology.

The inventor, however, knew in his heart that his technology was capable of substantially improving rolling resistance and that this difference could improve the fuel efficiency of vehicles. He was convinced it would reduce the consumption of fuel and our dependency on foreign oil. It seemed like a noble cause, and he was undaunted.

He had made a variation of the invention applicable for auto racing. Reducing the rolling resistance of race tires could help win races. Top race teams often use the latest available technology to stay ahead of their competitors. Using racing as an entrée to prove new technology had helped other inventors introduce new technology to automobile manufacturers.

I worked with a business consultant to create a business plan for the inventor's technology. We worked on the premise that the invention would prove itself through the racing forum and so gain acceptance from engineers and others in the tire-manufacturing business. We approached a prominent businessman who was a self-made entrepreneur and who had a racing team. It didn't hurt that he was also a golfing buddy of the chairman of Goodyear Tire and Rubber Company.

This businessman believed that he could use the technology first to his advantage for his racing team, then waltz into Goodyear to present the invention. He expected to see the technology implemented in production lines throughout the world. He committed $300,000 to the project over the next year and created a royalty structure for the inventor.

The only thing the inventor needed to do to complete the final stage of the technology for racing tires was to incorporate special high-speed bearings from a German manufacturer. Meanwhile, the businessman tried to use his influence to interest the chairman of Goodyear in the technology.

It never happened. The chairman would not circumvent his vice president of engineering to influence the decision of line engineers, and the engineers working in the trenches were already making improvements in the tire-manufacturing process. They were changing compounds to improve consistency in the mold, rather than fixing the tire after it came out. They were satisfied that by the time they had finished their research, it would preclude the need for the inventor's technology. In any event, it definitely would minimize its usefulness such that it wasn't worth pursuing.

A year went by, and the inventor was unable to get adequately performing bearings from Germany for the racing application. After eleven years of hard work, toil, and investments, the project died.

HINDSIGHT LESSONS

It would be easy to look at this project from the outside and conclude that the conspiracy between the oil companies and the auto manufacturers had yet again squashed a great technological improvement. This sort of thinking helps to support the fears of paranoid inventors. Even people close to the project wanted to blame the rejection on conspiracy, rather than look at the facts.

Disappointment aside, what could have been done differently? The project's timing could have been improved. Since AS Co. never followed up with the inventor, he concluded that they were not interested in his technology. Had the inventor not made such a misinformed conclusion, AS Co. might have helped co-develop the project five years sooner. Tire industry engineers would have been much more receptive at that time. There was a short window of opportunity to introduce this technology. The inventor missed it.

From the inventor's technical standpoint, and from the standpoint of people who worked for him at his laboratory, the invention was clearly a success. They were able to achieve what they set out to achieve, which was to improve rolling resistance.

From an investor's standpoint, however, the project was obviously not successful. Not only did they not achieve a return on their investment, they lost everything except for the pittance of revenues that came in for the licensing option.

One of the investors secured his $600,000 with a bank note contingent on his personal signature. His quarterly interest payments were nearly as high as the option payment. Needless to say, this investment was financially eating him alive. Luckily he was in a position to afford it. Why in the world would any investor assume debt liability for such a high-risk project? In this case, the investor was overly optimistic about the potential result of his project. He, like many people, thought that if this invention could save the country millions of barrels of oil, it would be a sure winner.

It's clear why an investor would normally want to make an equity investment, such as stocks, rather than use a debt instrument such as a bank loan. With equity participation the investor would own a portion of the inventor's project, and when the project became successful, the investor would receive a predetermined share of the proceeds, such as dividends. On the other hand, in the event the project went bust, the investor would simply lose the original investment, and not the ongoing interest charges associated with a debt loan. This is also why you do not find banks and other such lending institutions investing in high-risk projects in general, let alone inventions. Banks need to have collateral or net worth in support of the loan so that if the loan fails, they can foreclose and

WORDS OF WISDOM

Why should one or two investors bear the entire burden of risk on a project? Although the investors stood a remote chance of financially gaining from this project, society would ultimately benefit from this technology. Shouldn't society bear the burden of risk? This may be so in a socialist government, yet many inventors believe that the federal government should invest in independent inventors' research and development. Many inventors also assume that it is the Patent Office's responsibility to commercialize inventions. This is definitely not the case. The Patent Office has no way of knowing whether an invention will benefit anyone, let alone have any degree of commercial success.

get their money back, or at least a portion thereof. Not to mention that banks want to get the interest, too.

Some inventors choose to mortgage their house to pay for the development of their invention; this is a risk that every individual must weigh. It should be looked at as if you were going to Las Vegas. You would not want to risk any more time and money than you are willing to lose.

Every now and again inventors approach me with stories about how they were laid off from their jobs, are running out of unemployment, and are relying on this idea they dreamed up to save them financially before the checks stop. When I hear this, I want to suggest that the chances of being struck by lightning are better. However, being the eternal inventor optimist, I never rule out the possibility that inventors will luck out and present their idea to the right company at the right time and earn royalties that put them on Easy Street. Improbable—highly improbable—but not actually impossible. The problem is that so many inventors, with even the most marginal of ideas and inventions, feel that this will happen to them.

Most inventors produce inventions that work well, and they can prove it. They can even show how their inventions will save time and money. But this is not enough. You have to consider all the factors (and all the people) that will influence your project.

Determining the factors that will influence the commercialization of your invention is like sleuthing. One of the best techniques I have seen is displayed by the television detective Columbo. It simply requires that you humbly ask a whole lot of people a whole lot of questions, listening well and saying no more than is necessary to keep the other person talking. It is extremely important to gather research in the initial stages of your invention project. To do otherwise is like diving off a high board without knowing the water's depth.

Inventions that don't succeed, like seeds that don't grow, are not missed.

—ANONYMOUS

CHAPTER TWO

Timing Is (Almost) Everything

Formulating Your Patent Strategy and Protecting What's Yours

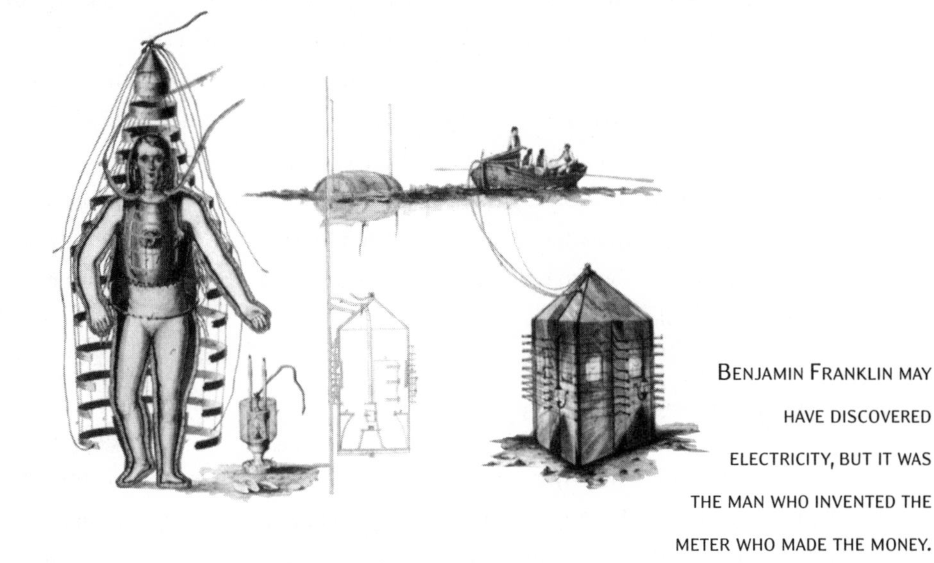

Benjamin Franklin may have discovered electricity, but it was the man who invented the meter who made the money.

—Earl Wilson

Many inventors believe that patenting their invention is a necessary first step to licensing and selling it. Many also think that once they have a patent, their invention is protected. Both of these beliefs are true to a degree, but like so many things in life, it's more complicated than that.

As we saw in Chapter 1, patenting isn't always the best first step in the commercialization process. Patent applications are expensive and time consuming, and market and industry research often reveal that an invention won't sell enough to warrant a patent application.

Whether or not you patent your invention early on partly depends on your budget. It also depends on where a patent application ranks in your overall commercialization strategy. At some point in the process—usually around the time you're getting ready to submit your invention to potential licensees—you'll need to start the patent ball rolling. We'll get to that in later chapters. First, though, you need to do preliminary patent searches, start protecting your proprietary property, and develop a patent strategy for when you do hold a patent.

Diving dress by Chauncey Hall (U.S. Patent No. 1405X), Connecticut, 1810.

Just What Is a Patent?

A U.S. patent for an invention is the grant of a property right to the inventor(s), issued by the U.S. Patent and Trademark Office (USPTO). The right conferred by the patent grant is, in the language of the statute and of the grant itself, "the right to exclude others from making, using, offering for sale, or selling" the invention in the United States, its territories, and its possessions, or "importing" the invention into the United States, beginning with the date of patent grant and extending for a period of twenty years from the date of initial application. To get a U.S. patent, an application must be filed in the USPTO. Maintenance fees must be paid to the USPTO every few years to keep the patent valid. Design patents for ornamental devices are granted for fourteen years from the date of issue.

Appendix A contains an overview of the U.S. Patent and Trademark Office's (USPTO) brochure on patents. This will familiarize you with the nuts and bolts of patenting, including what can and can't be patented, the application process, and common problems that patent seekers run into.

How Necessary Is a Patent?

A brochure published by the American Patent Law Association suggests that any idea is useless without patent protection. Obviously this organization is interested in supporting the livelihood of its members: patent attorneys. Their statement regarding patent protection is true in most circumstances, but not all.

For example, I recently spoke with the president of a company that manufactures hand tools. They had paid royalties to twelve independent inventors, six of whose ideas were unpatented. Patents are a tool (pun intended), not an end result. If inventors truly understood the nature of patent claims and their relationship to commercialization, there would be many fewer patent applications being filed. A patent can be so limited in its scope that it's essentially useless. (More about this later in this chapter.)

Safeguarding Your Proprietary Property

It is of utmost importance to accurately document the conception date of your invention and any improvements thereafter. This does not afford you patent protection; however, it does establish the date of conception. If another person conceives the same invention, it may be necessary to prove that you did not derive the invention from another person who filed a U.S. patent application. Although such derivation proceedings at the USPTO may be rare under the new patent laws, there may nonetheless be circumstances where proving your original inventorship may be imperative.

Another requirement for receiving a patent is that you must have "reduced your invention to practice." Examples of "reduced to practice" include making a working model or filing a patent application. Essentially, you can file for a patent prior to making a working model when the basic concept of your invention is readily accepted by known science.

Protection starts with timely documentation.

AMERICA INVENTS ACT (AIA) REVISED PATENT LAW

The America Invents Act (AIA), effective March 16, 2013, changed the U.S. patent system from a "first to invent" system to a "first inventor to file" system. No longer does the inventor who first conceives an invention have first shot at a patent. It's the first inventor who files the patent application who will have priority. Most of the world's developed nations have been on the first to file system for decades. The U.S. historically resisted changing to the first to file system, since the U.S. Patent Office was created in part to acknowledge original inventorship.

Now that the U.S. has embraced the "first *inventor* to file" system, you may no longer have priority by being the first inventor to conceive an invention. You nonetheless have to swear that you are an inventor or conceiver of the invention, and that means NOT plagiarizing it from another person or source. Yet, there was fear that the first to file system may allow little in the way of safeguards to prevent unscrupulous individuals from claiming an invention is theirs, when indeed they really learned about it from another.

WORDS OF WISDOM

Your patent application will be published publicly in the *Official Gazette* of the USPTO, unless you elect to not have it published, which you must do when you file your original patent application. See Appendix A for more information.

This new system merely underscores the importance of being diligent about disclosing your invention only on a need-to-know basis, and only to trusted people, especially if you are doing so prior to filing your patent application.

More about what constitutes appropriate justification for filing a patent application, and when and to whom it may be best to disclose your invention, is found later in this book.

PUBLIC DISCLOSURE OF INFORMATION

Inventors in the United States have a one-year limit, called a *grace period,* within which to file a utility patent application after their invention has been on sale or offered for sale, is in public use, or is patented or described in a printed publication in this or a foreign country. It is extremely important to consult a patent professional about this issue.

Public disclosure starts the patent clock ticking.

After one year the inventor is time barred and loses their right to file a patent application. Your invention is then available free for public use. The decision to publicly disclose your invention is an important part of your strategy. If you publicly disclose your invention you could lose the chance for some foreign patents or be forced to apply for a patent before you want to do so.

Inventors are also reminded that any public use or offer for sale of the invention—or its publication anywhere in the world—more than one year prior to filing a patent application for that invention will prohibit the grant of a U.S. patent. Foreign patent laws in this regard may be much more restrictive than U.S. laws.

The issuance of a U.S. patent is considered a public disclosure and could bar you from applying for foreign patents, especially if you have not filed under the Patent Cooperation Treaty (PCT). Filing for a foreign patent gets complicated and expensive ($20,000 plus), so use a patent professional for this.

If your invention has strong potential in foreign markets or you feel there's a lot riding on your patent process, it is of utmost importance to consult with a patent professional *before* you publicly disclose your invention, offer it for sale, or use it for your own commerce, even if kept confidential. Factors such as these may affect your potential to obtain patent rights.

Patent Searches

Normally the first prudent step in the patenting process is to have a patent search performed for you or to perform an informal patent search yourself. These searches are also referred to as *prior art searches.* The USPTO relies on industry publications and patents to establish prior "art." Having a patent attorney collect prior art information and render a patentability opinion will cost $500 to $1,500. Some firms will use a patent agent to perform searches and offer a patentability opinion for half this fee. The Resources section lists several resources for finding patent professionals. Doing it yourself can cost from nothing to $100 depending on the method you use.

PRIOR ART

Patent searches are sometimes called *prior art searches.* When patent examiners at the USPTO review a patent application, they search prior world patents, published patent applications, brochures, the Internet, and other materials. Any of these sources can reveal a similar product and cause an application to fail.

How extensive should your patent search be? A lot depends on how much is riding on your invention, and how important patenting is in your overall commercialization strategy.

If you have a simple invention, or if your invention is in a niche market where you would not anticipate hundreds of thousands of dollars in royalties, you may want to consider using a patent agent or a do-it-yourself search. Follow this with a market investigation as described in Chapters 3 and 4 to determine your invention's marketability before spending much more money on the patenting process. This strategy will help keep you from paying more money for developing your invention than you will earn back.

On the other hand, if you have a hallmark invention, like a formula that will reduce our dependency on fossil fuels by 10 percent, you should enlist the services of the best patent law firm in your field. This won't come cheap. Expect to pay hourly rates of $200 or more. See Chapter 7 for more about choosing and working with a patent attorney.

It took 120 years to register the first million patents, but only 3.5 years to register the most recent million.

BENEFITS OF PATENT SEARCHES

Preliminary patent and prior art searching is a critical part of the commercialization process because it shows whether someone owns a patent on your invention, even if you have never seen it in the marketplace. Remember that less than 2 percent of all patents are ever commercialized, and fewer than this make enough money to pay for initial expenses. You certainly don't want to pursue an invention only to find out that someone already owns the patent rights. I see this all the time after inventors return to me with their patent search results, surprised to have learned that their brainstorm has already been patented.

Even if the results from your search don't reveal another invention similar to yours, it doesn't mean your invention is patentable. There may be a patent in your general subject area that remotely resembles an element of your invention. There may be other patents not directly related to your invention that possess individual elements found in your invention. When you combine all these features and elements together, you end up with something close to your invention.

If this is the case, the examiner who reviews your patent application may decide that it would be obvious to someone skilled in your art to combine these various elements if they knew they existed. This is a hypothetical argument, but this one argument alone, the "obviousness rule," is cited in the majority of patent rejections at the USPTO.

It is because of these intricacies that a thorough patent search may be particularly important. It can be helpful to

From One Inventor to Another—A preliminary patent search identifies other inventors in your field. These inventors may have prior experiences with introducing new products in your market. You can call these inventors and ask them how their invention has progressed in the marketplace. I have done this on numerous occasions and have found that inventors are always quick to share their trials and tribulations. They can tell you about complications in the marketplace that you may not have considered. The more you learn, the better you are able to anticipate these challenges and possible barriers.

use a professional with a lot of experience in all aspects of the patent process who will be better able to recognize the elements in prior art that may affect the patentability of your invention. This is another example of why the cost of patent searching varies so much. In one respect, you may get what you pay for, although an expensive search does not always equal a better search. I believe there is no substitute for an experienced patent professional, regardless of the cost involved.

There is additional value to patent searches even when your invention is novel and patentable. A patent search reveals a tremendous amount of information about other developments in your field, which can give you ideas about improving your invention. There may be other patents covering similar concepts, but not exactly like yours. You can refine your patent application so that it does not interfere with others' patents.

FREEDOM-TO-USE OR FREEDOM-TO-OPERATE

Even when you have been issued a patent, your patented invention may infringe another inventor's patent. Remember, a patent gives the right to exclude others, not the right to practice your invention.

For example, let's say that Edison patented the first light bulb, and his patent claimed a light-emitting device that utilizes electricity. The key words are "light-emitting" and "electricity," and these in combination are what he is excluding others from using.

Thereafter, you may come along with a brilliant improvement to make the light shine longer, burning in a vacuum with argon gas. These may be new features that are patentable, yet you must rely on Thomas's invention in order to practice yours. While his patent is valid, you will need his permission or a license from him.

In order to know if your invention may infringe another, you may obtain a freedom-to-use (or freedom-to-operate) legal opinion from your patent attorney. The USPTO does not issue these opinions, and it has no jurisdiction over questions of infringement and the enforcement of patents.

The cost for a freedom-to-use opinion may be five to ten times that of a patentability opinion because it involves an extensive patent review.

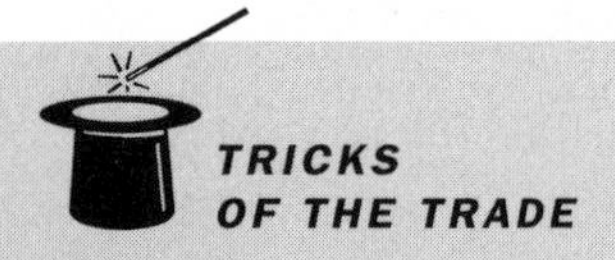

TRICKS OF THE TRADE

Assignee Searches—Patents are listed under the names of the individuals who conceived the invention. When an inventor works for a corporation, that corporation often owns the patent rights based on prior agreements between the company and the individual. In this case the company will be listed as the *assignee* of the invention. The assignee listing is found on the first page of the patent directly under the names of the individuals. When your patent search reveals other patents in your field, you can see which corporations might have an interest in inventions like yours. You can perform a separate *assignee search* to discover all the patents assigned to companies in your field.

A compromise may be to ask your patent attorney for a freedom-to-operate "comment" only for the patents found in your prior art search for patentability. It is better than no search at all and a lot less expensive.

TRICKS OF THE TRADE

Patent Classes—Starting in 2015, the USPTO Office of Patent Classification harmonized the U.S. and European systems to form the Cooperative Patent Classification (CPC) effort. The U.S. Patent Classification (USPC) system of over 400 classes and 150,000 subclasses is still maintained. Now, for thorough results, a search of *both* systems is required, in addition to other world publications for an exhaustive review of the prior art.

U.S. PATENT AND TRADEMARK RESOURCE CENTERS (PTRC)

There are over eighty resource centers, formerly called Depository Libraries, throughout the United States and its territories. They are usually located in public or university libraries in metropolitan areas. These resource centers have most of the over eight million patents on DVD, CD-ROM, or microfilm dating back to the origin of the patent system. These centers are free and are the next best thing to visiting the USPTO search room in Alexandria, Virginia. Most resource center libraries have detailed information about how to perform patent searches, and their librarians are trained to show you how to conduct patent searches at no charge to you. Call ahead to be sure a librarian will be available to help you. For online help go to: http://uspto.gov/products/library.

Other PTRC services and resources vary between libraries and may include:

- CASSIS, a CD-ROM search indexing tool for patents and trademarks.
- Online search systems of the USPTO. They feature full patent images from 1790 to the present.
- Espacenet, allows searching and printing of European patent applications and issued patents. Also available through: www.epo.org.
- TESS, an online database of the USPTO that allows searching by text or design of federally registered trademarks.
- Directory of attorneys and agents registered to practice with the USPTO.
- The *Official Gazette,* the weekly publication of the USPTO, listing all newly issued patents and trademarks, new patent rules, and other information.
- Periodic seminars and conferences relating to intellectual property issues.
- Document delivery of U.S. patents, international patents, file wrappers, and assignment records.
- Sale of government publications and popular materials on patents and trademarks.

PTRC resources can be used to:

- Determine whether an idea may be already patented
- Define specifications and drawings of known patents
- Obtain state-of-the-art information on specific subject areas
- Prepare a list of prior developments to be cited on your patent application
- Obtain patent history of a specific inventor or company
- Research inventions and inventors of historical or personal interest
- Access information about registered and pending federal trademarks
- Obtain information about applying for patents and trademarks

The USPTO website has several great resources for patent searching and other help. For searching, see www.uspto.gov/patents/process/search/index.jsp. For a step-by-step tutorial of the entire searching process, see www.uspto.gov/video/cbt/ptrcsearching.

HOW TO PERFORM A PATENT SEARCH

Here's my method for searching in a PTRC or online. First, I do a key word computer search to identify patents in my general area of interest. I note the class and subclass of similar inventions. I cross-reference these results with subclasses found in the *Index to the U.S. Patent Classification System* and CPC. The index is a point of entry to the *Manual of Classification* which provides a fast read of the patent classes and subclasses. Once I narrow down the possible subclasses, and there may be several seemingly appropriate ones to choose from, I proceed to the *Patent Classification Definitions*. Reading through the definitions, I pick those subclasses that best describe my invention. The *CASSIS* CD-ROM program lists all patents in a given subclass, usually fifty to a hundred. From this list I either refer to the *Official Gazette of the USPTO*, which has a one-page abstract and a drawing of all patents listed by patent number, or I refer to the USPTO computer database, or www.google.com/patents.

From this initial review I determine if I am indeed in the right subclass based on the patents I find. You really need to go through darn near all the patents in the subclass to be sure. As a result I may refine my search by looking at other definitions, or go ahead and pick those patents that look most interesting and search *CASSIS* or the best available database to view the full text of the patent and all its drawings. Search instructions may also be found in the introduction of the *Manual of Classification* and in the *CASSIS Reference Manual*.

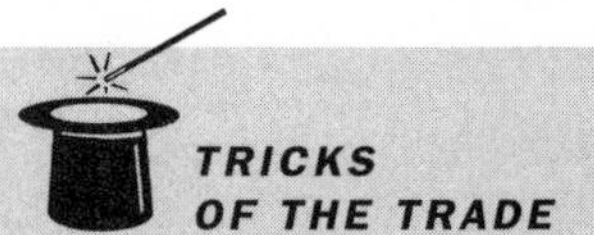

Patentable Advice—You may want to take the results of your preliminary patent search to a local patent attorney or two for free initial consultations. The patent attorney will review the patents that you have found and determine whether your invention is too similar to the other patents to be patentable.

After doing a few of these searches, the concept of hiring a professional to do the search seems much more compelling.

Like online searches, resource center patent searches should be considered as preliminary searches because they are not as thorough as a search performed by a professional at the USPTO, nor will they reveal pertinent industry publications. They are, however, a good first step to help you learn about similar patents.

U.S. PATENT AND TRADEMARK OFFICE SEARCH

The USPTO search room is open to the public, and you can perform your own search there. The procedure is similar to the one described for the U.S. patent depository libraries. The search room is located at Madison East, First Floor, 600 Dulany Street, Alexandria, Virginia 22314. (Phone: 571-272-3275.)

There you will find state-of-the-art computer workstations that provide automated searching of patents issued from 1790 to the current week of issue. Full document text may be searched on U.S. patents issued from 1971 to the present. Patent images may be retrieved for viewing or printing.

No two patent searches are ever the same.

Prior to filing a patent application, it would be wise to invest in an exhaustive search done by the patent attorney or agent who will prosecute your patent application. Your preliminary search could overlook a patent that could cause your patent application to be disallowed. Investing in an exhaustive search by a professional increases the odds that you'll find prior art before you apply. Why spend a few thousand dollars on a patent application when for only a few hundred dollars you can determine the viability of your application in its current form?

CONTACTING THE USPTO AND PTDL

See the listings for U.S. Patent and Trademark Office (page 235) and U.S. Patent Depository Libraries (Appendix B) for their addresses and phone numbers.

There is no method of patent searching without room for error. Even a professional search can miss a patent reference. I have never seen two patent searches for the same invention turn up exactly the same results. This is a harsh and sometimes disappointing reality of the patenting process.

Provisional Applications

The USPTO has provisional applications for patents that help small independent inventors. For about $500 you can file your own provisional application (or have a patent professional file one for you at a cost of up to $1,500 or more). The advantage is that you can inexpensively have an application on file, and it establishes your priority date. It is less likely that public disclosure of your invention within twelve months of your date of priority will adversely affect any foreign patent applications. You do not have to submit formal drawings or claims; however, you must fully disclose all of the pertinent elements of your invention for it to be later converted into a regular patent application.

Within one year from when you filed your provisional application, you may then elect to convert your provisional application into a regular utility patent application. Basically, this buys you one year before having to make a larger expenditure. During this time you can learn more about the marketability and licensability of your invention. If you find that your invention is not worth pursuing further, you have less money at risk.

The down side of this is that if you decide to go ahead and file the utility patent application, you will likely incur as much cost to convert the provisional application as you would have paid had you filed the utility application first. You are ultimately paying more. This gets into a very complicated area, which is why many patent professionals advise against filing provisional applications. The end costs may outweigh the benefits.

Many independent inventors prefer making provisional applications on their own because it is inexpensive and it gives them a measure of comfort. It also gives your invention a patent pending status when you approach major corporations. As we'll see in a later chapter, having a patent pending status may be necessary in submitting to some companies, especially big ones.

Shop Rights for Employers and Institutions

If you conceived or worked on your invention during employment hours, or used your employer's resources, such as a lathe or a computer, your employer may have the right to insist that any patent rights arising from your invention be automatically assigned to the company. This may be true even if your invention has nothing to do with the type of product or service offered by the company. The laws governing an employer's right of ownership to an employee's patent or other proprietary property vary from state to state.

If you are hired as an engineer to develop products for your company, your employer may also have rights to your invention even if it is developed off the company's premise or outside of employment hours. Most science and engineering firms or engineering departments of large companies have contracts with engineers that clearly outline each party's specific rights. Typically, employees relinquish all proprietary rights to the company for a dollar and "other good and valuable consideration," like their salary.

Generally speaking, if you are not using any of your employer's resources, are working outside employment hours, and are not inventing in a field related to your company, it is likely that your employer cannot assert any rights to your invention without a written agreement stating otherwise. Consult a patent attorney in your state who specializes in employer/employee relationships for further information about this.

When employees have a very restrictive agreement with their employer regarding intellectual property, it has been my experience that most corporations will waive their rights to the employee's invention if that invention has no relationship to the company's business, even when the employee uses some of the employer's resources. Given the choice, companies prefer to have a more restrictive contract with an employee with the option to waive those rights than vice versa.

INTRAPRENEURSHIP

Some innovative corporations have provided a safe haven for employees by offering them the opportunity to start their own company (or profit center) under the umbrella of the parent company. The term for this is *intrapreneurship*. The concept of intrapreneurship has not yet caught on because, as you might expect, a startup endeavor is a risk, whether or not it is associated with a large company. The employee needs to be a proficient entrepreneur and to create a viable business plan, and the employer must be flexible, insightful, and willing to risk. This certainly doesn't describe a majority of employers, but it may be worth your while to suggest an intrapreneurship arrangement with your employer.

WORDS OF WISDOM

To learn more about the value of patents in the corporate world, check out *Rembrandts in the Attic: Unlocking the Hidden Potential of Patents* by Kevin G. Rivette and David Kline (Harvard Business School Press, 1999). Use the information as an ace in your pocket when pitching the concept of intrapreneurship to your boss.

UNIVERSITY INVENTIONS

Universities and colleges usually have special contractual relationships with their professors, teachers, and graduate students relating to shop rights. Universities often insist that any intellectual property developed be assigned to the university. The technology transfer department of the university decides whether or not a patent application (or other efforts) should be filed for licensing or commercialization. Inventors may have the opportunity to pursue an invention on their own if the university has decided not to. In this case it would be up to the inventor to pay for additional development and patenting costs and arrange for commercialization.

There are licensing and technology transfer organizations that deal specifically with smaller colleges and universities that cannot afford their own technology transfer departments. They work as agents in getting the invention licensed, purchased, or co-ventured with an outside company. These technology transfer organizations usually have a commission relationship with the university and a specific agreement defining who will pay what with respect to patenting and other development costs.

Many universities have earned tremendous royalty income from their employees' inventions. Some institutions consider inventions a potential income base and encourage inventors. University technology transfer agents are usually quite accessible and willing to talk shop with an independent inventor. It could be worth your while to do so.

GOVERNMENT RIGHTS

Government employees and people who use federal grant money now have a much easier time maintaining the rights to their intellectual property. Government employees, the military, and people who received federal grants used to be required to assign all their invention rights, title, and interest to the U.S. government, royalty free. When the government decided to encourage the transfer of technology from the military to the private sector, the administration loosened its intellectual property policies.

Now the federal government routinely offers its employees and grantees an opportunity to maintain the rights to their inventions so they may pursue commercialization in the private sector. The government does maintain the right to exploit inventions within the realm of the government; however, outside that realm inventors are free to do as they wish. Arrangements vary, though, and you may find different stipulations depending on the area of government you are working with or the type of grant you received. The federal government has the right to exploit your patent, but if it does, it must pay you fairly.

Your patent rights may belong to others without your knowledge.

Patent Partners

If anyone besides you contributed significantly to your invention, they will likely need to be included as inventors on your patent application. Always consult a patent professional about this issue. When there is a co-inventor, in the absence of any written agreement to suggest otherwise, you will own an undivided interest in the invention. This means that although you both may have a fifty-fifty stake, either one of you may sell all or part of the patent rights without the knowledge or consent of the other. To avoid a potentially sticky situation, be sure to draw up an agreement between you and your co-inventor that clearly outlines ownership and the right to exploit your invention. An once of prevention is worth a pound of lawsuit. For more information about this, see Appendix A.

SELECTING A PATENT PROFESSIONAL

Patent attorneys are lawyers who are registered with the U.S. Patent and Trademark Office (USPTO) and who have sworn to practice according to the rules and regulations of the USPTO. There are nearly twenty thousand patent attorneys in this country, and they are generally trustworthy. If it is discovered that they have treated their clients improperly, they can be disbarred or disallowed from practicing in the USPTO. Needless to say, they have more to lose than gain from stealing an invention, so this isn't something you should worry about.

A listing of patent attorneys can be found in the Martindale-Hubbell directory (available online and in most major libraries). The USPTO also publishes a list of its registered patent attorneys and agents. Most patent attorneys work exclusively for corporations; however, a selective few represent independent inventors. I find the best ways to identify those who will work with independent inventors is to search the yellow pages under Patent Attorneys, or get a referral from a local inventor's group. Information about these groups is found in the Resources section.

A patent attorney's hourly rate generally ranges from $150 to over $300. Some law firms are comprised of a group of patent attorneys. The larger firms have patent attorneys specializing in specific areas of expertise, such as electrical, chemical, biotech, computer software, trademarks, and so on. If you are inventing in a specialized area, you may want to seek out a patent attorney who knows that field.

Some patent attorneys specialize in particular aspects of the patent process, such as filing patent applications, licensing, or litigation of patent lawsuits. I prefer to work with attorneys who handle all of these aspects; their broad base of experience brings greater depth to the application process and helps me anticipate potential licensing contract issues and hopefully avoid the need for potential litigation.

Every tool carries with it the spirit by which it had been created.

—Werner Karl Heisenberg

PATENT AGENTS

Another group of individuals authorized to file patent applications and prosecute them with the USPTO are patent agents. They are not attorneys, but they do practice with the USPTO. Their fees tend to be lower than those of patent attorneys, and they generally price their services per job or per a particular function of their job, rather than by the hour. For example, they may have a flat rate for filing a patent application ($1,500 to $3,000) and another rate for handling other aspects of the patent process, such as interviewing with patent examiners, appealing the patent, and so on.

Hiring a patent agent does not mean that you are going to receive inferior service. I know patent agents who have obtained patents for inventors when a high-priced patent attorney failed to do so. Some patent agents are former patent examiners of the USPTO. Such experience can go a long way toward helping you.

WHICH PATENT PROFESSIONAL SHOULD I USE?

At the beginning of the patent process, I like to have a Washington, D.C., patent agent firm do preliminary patent research. Because they can do their searches at the USPTO, their results can be more thorough than those done strictly via computer database.

Next, I like to sit down with two or three local patent attorneys who are already doing business in the area of my invention and get their opinions regarding patentability and commitment to the project. The initial consultation is normally free and can help you decide whom to hire to do your patent work. Although a large patent firm may have attorneys with greater experience, they may assign you to a new associate with little experience. It is important to know who will be doing the actual work on your patent application.

WORDS OF WISDOM

The key elements involved in the selection of a patent professional are:

- type of protection needed
- their level of expertise
- profit potential
- budget
- personality mix

It is also important to know the degree of experience the firm has with filing foreign patent applications. Normally a U.S. patent attorney, and in some instances a patent agent, will get the foreign patent ball rolling by filing internationally under the Patent Cooperation Treaty (PCT), the Paris Convention for the Protection of Industrial Property, or by filing with the individual country. This is often a complicated, cumbersome, expensive, time-consuming process and normally involves hiring patent attorneys or agents in individual countries and in some cases hiring a translator. A good way to determine if foreign patenting fits into your long-term commercialization strategy is to follow the techniques described in the ensuing chapters about how to gather the market information necessary to make informed decisions without publicly revealing your invention.

Some Strategies for Patent Holders

Whether you are at the idea stage or have a patent in hand, it is important to protect your invention from being copied in a way that destroys its licensability. A patent is only as good as the claims you make in the claims section of your patent.

PATENT CLAIMS AND COMPETITIVE ADVANTAGE

Prepare your strategy for proprietary protection carefully! You may be entering a very crowded and competitive marketplace where what you have to offer boils down to the specific claims of your patent. The claims outline exactly what you are excluding others from doing. The detailed description of your invention contained in the patent teaches an observer how to interpret the claims. These claims and the "teaching" in the text of your patent are used to determine whether or not someone is infringing on your patent. The claims shouldn't be so limited that others can circumvent your patent and design a product with strikingly similar features.

Some companies (and some entire industries) will only honor your specific patent claims, and with good reason. If they can easily produce your invention without using your patent, why should they license it and pay you royalties? Before approaching potential licensees, you need to understand the strengths and weaknesses of your offer to them. If you discover that your patent is indeed narrow in scope, contact potential licensees that have a track record of recognizing your total license package (see "What Do You Have That You Can License" in Chapter 1).

A patent is only as good as its claims.

WHAT'S IN A CLAIM?

A good patent attorney or invention developer will examine your patent's claims for possible weak spots. If you're commercializing your invention on your own, you need to do this too.

Each of your claims will contain one or more specific elements, and the number of claims you have depends on the complexity of your invention. Here is an example of an apparatus claim in a utility patent for a padlock.

I claim:

A padlock comprising in combination:

> A shackle having a generally U-shaped configuration with one short leg and one long leg;
>
> Said long leg having at a free end thereof a flange which underlies a cylindrical portion which in turn is spaced below a notch on said long leg;
>
> A padlock main body having an opening therein to receive said shackle long leg, and another opening therein to receive the shackle short leg, said padlock main body further having a blind bore including an open end formed therein;
>
> Lock means within said main body to engage the said long leg either at said notch, thereby locking said long leg, or to engage only the long leg at said cylindrical portion, thereby allowing said long leg to rotate about its longitudinal axis, or to removably free said long leg by having said lock means clear said flange, thus defining three positions;
>
> A stop means disposed in the blind bore of the main body between the lock means and the open end of the blind bore, whereby the lock means engages and cooperates with the stop means preventing accidental removal of the lock means from the main body.

This claim has about six key elements. Basically, each paragraph is a different element, starting with the fact that we are dealing with padlocks and then including a combination of additional elements. For people to infringe on this patent, they would have to infringe on *all* of these key elements. If they designed an apparatus that did not include one or more of these key elements, they would likely not infringe on the patent.

Therefore, one of the first questions I ask clients is: "Could anyone design a product that would eliminate a key element and still obtain the desired result?" In the above padlock claim, the question might be: "Could this padlock be designed to work if you eliminate the sixth element (for example), 'a stop means . . .,' from the design?" If the answer is yes, then there is questionable value to the patent. The reason for this is simple. Any company that licenses your invention will be at a disadvantage against its competition. Without infringing on your patent, the competition can copy your technology, change one key element, and never have to pay a licensing fee. They can undercut your licensee's pricing because they are not paying royalties to you.

On the other hand, if someone were to design a product that included all of the key elements of your claim and more, this is likely an infringement on your patent since they incorporate *at least* all of your key elements.

A method claim is similar to one that describes an apparatus, and both can be included in a utility patent. However, a method claim describes *how* your invention operates, not what your invention is composed of. Method claims may cover a broader range of activity. The broader your claim, the broader your protection. Optimally you would start with one or more broad claims, then add additional claims for all possible variations.

Foreign Patenting

Because of the important role that patents can play in foreign markets, it is important to consider the potential for foreign patents from the outset. Most industrialized countries will not permit you to apply for a patent after your invention has been publicly disclosed. A public disclosure can come from an article in a trade journal, sale of your invention to the public, or the issuance of a U.S. patent. Therefore if you see substantial foreign markets it may be in your best interest to make sure that your invention is not publicly disclosed while you are identifying potential licensees.

This is one reason (we'll see more in the next few chapters) why it is so important to be selective about the companies to which you disclose your invention, and to pay attention to the reputations of companies with which you communicate.

Bear in mind that it does not matter who releases the information about your invention. A public disclosure can come from any source and may adversely affect you.

THINGS TO REMEMBER AND CONSIDER

- Perform a patent and prior art search.
- Identify limitations that may encumber your rights of ownership.
- Develop a rapport with a patent professional.
- Understand how patents and other rights can work for you.

TRADE DRESS

Trade dress is a form of intellectual property when used as a secondary meaning to a trademark. It refers to the characteristics of the visual appearance of a business (such as the design of a building); a product's unique shape, color, or packaging; the design of a magazine cover; or other distinguishing nonfunctional elements that signify the source of the product to consumers.

CASE STUDY

An Automotive Accessory (Part 2)

In the first segment of this story, things seemed to be going pretty smoothly. My invention was successfully licensed to a major manufacturer, and the patent was pending with the USPTO. Here's what happened next . . .

After Rubber Queen made a market introduction splash and started distributing the mirrors, they realized they would have to price the mirrors less than they had anticipated to garner adequate sales. With my royalty rate very high, they couldn't be competitive with the other stick-on spot mirrors on the market. I agreed to reduce my royalty rate per unit in half, in exchange for three lump payments totaling $25,000 as an advance on future (reduced) royalties.

Then the U.S. patent that I had originally applied for was finally rejected after an extensive appeals process. The whole patent process had taken nearly four years. This was something of a freak circumstance because the patent actually had claims allowed, and we were waiting for it to issue when the rejection was announced. As it turns out, our patent examiner believed there were too many patents on the market and felt that an inventor needed to have an exceptional improvement to be awarded a U.S. patent. He argued that it would be obvious for someone skilled in the art to create the rectangular wedge mirror, even though nothing else had been brought forward that looked anything like it. This is a good example of an "obviousness" rejection. The examiner found additional prior art and the patent was rejected as a "final" with no other recourse. I think it was ultimately the patent examiner's personal opinion that turned the tide on my patent. That's part of the patenting game. Your patent is one chicken that you definitely don't count until after it is hatched and moving. Even then, there are predators.

Rubber Queen decided to scrap the Docie wedge mirror project. The patent was rejected, and Rubber Queen knew that some of its competitors were poised to take advantage of this situation and introduce competitive products. Several manufacturers we had originally approached had liked the product but preferred to see whether the patent did indeed issue before they licensed with me. When they learned that it didn't issue, they were ready to pounce. Two or three manufacturers started importing wedge mirrors with designs very close to mine and distributing them on a mass basis. They were smart enough to slightly change the design of my mirror just in case I had a design patent lurking in the bowels of the Patent Office.

In fact, my investor/partner did have certain design and utility patents pending in the United States and other countries for improvements to my original concept. The problem was that none of these had issued, nor did we anticipate them issuing in the near future. All we could do was watch the Docie wedge mirror progress in the marketplace with growing distribution and with no royalties paid to anyone.

Over the next five years, due to the product being royalty free and because of greater distribution, companies lowered the price for each mirror from around $2 down to the $1 range. It hurt me to walk into auto parts stores and see my invention on the shelf, knowing that I wasn't getting anything for it. What hurt even more was to see the quality and design of the unit go downhill fast. It is a safety product that can potentially save lives. It needs to work properly. But the overseas manufacturers and the companies that were importing and distributing the mirrors saw them as just another $1 commodity item and didn't much care whether the angle of the wedge was enough to adequately cover the blind spot or not.

One day I was in a local auto parts store and noticed my mirror on the shelf. The quality was lower than low. The mirror's distortion was terrible. I said to myself, "The consumer is really getting ripped here," and I decided that I would see that this was remedied whether I was getting a penny for the product or not.

My first step was to talk to my former investment partner, Alex, who to my surprise had received a U.S. utility patent on an improvement to my mirror that would lower the manufacturing cost. It seemed to me that the current mirrors on the market might infringe on his improvement patent, and a notable patent attorney agreed with this conclusion.

The Automotive Parts and Accessory Association (APAA) trade show was in a couple of weeks, so I decided to travel to Chicago to meet with manufacturers at the show. Before going, I offered to purchase Alex's patent for $25,000 in payments to him. The payments were to be a portion of the royalties that I would receive if I concluded any deals with the manufacturers. Alex gave me a sixty-day option period to conclude the deal with him, which gave me enough time to approach the manufacturers. This was the start of what would turn out to be two years of extensive wheeling and dealing.

(To be continued in Chapter 3 . . .)

HINDSIGHT LESSONS

Being willing to improve upon my original design concept made a difference and ultimately kept the patent—and the product—alive. There have been about five distinct generations of the Docie mirror. In the beginning I thought I had the final product, the one that would make me rich. Now the improvements to my invention have the greatest promise, not the original design. To be successful, I would have to leave my ego behind and embrace someone else's design.

It is easy for inventors to be caught in the trap of not wanting to modify their inventions. Sometimes your original concept is integral to your invention. Other times you must move into new generations. Either way it is important to be open to market feedback on your invention.

The status of my patent played a significant role in this project. Having the original patent pending, even though it was eventually rejected, opened the door to my initially making money and getting my product into the marketplace. The improvement patent, as minimal as it seemed to be, provided the hope to potentially salvage this project. Neither patent position was ideal, so I would have to use other tactics to pull this rabbit out of the hat.

What one man can invent another can discover.

—Arthur Conan Doyle

CHAPTER THREE

Market and Industry Research

Your Ticket to Free Expert Advice and Lasting Business Relationships

THERE ARE THREE CATEGORIES OF PEOPLE IN INDUSTRY—THE FEW WHO MAKE THINGS HAPPEN, THE MANY WHO WATCH THINGS HAPPEN, AND THE OVERWHELMING MAJORITY WHO HAVE NO IDEA WHAT HAPPENED.

—O. A. Battista

You may know exactly what application your invention should be used for and exactly how it fits into a certain industry, which you already know everything about. You know which company you want to license to and whom you need to talk to there and what to say to them. You have a firm understanding of your invention's foreign and domestic markets and the distribution channels it will move through in each. You know all about the competing products, and you have ideas about the kinds of packaging, promotions, and shelf placement that will work for your product. You've figured out manufacturing costs and retail price and have used these figures to determine your potential royalty rate . . . or not.

If this doesn't quite describe you, read on. All this knowledge and more is out there, yours for the asking.

Just Talk to People

Interviewing the right people in an industry is at the heart of the commercialization strategy I follow as a successful inventor and as an invention development professional. It works!

Mower by W. Manning (U.S. Patent No. 6539X), Plainfield, New Jersey, 1831.

If you don't know the answer, ask someone who does.

Within any given industry you will find people who are more than willing to share their knowledge with you about everything from whether they see a place for your invention in the industry to how you can best present your invention to how they think others in the industry might perceive your invention. This chapter and the ones that follow will help you find those key people, interview them, and use their knowledge and expert opinions to help shape your commercialization plans. You'll emerge an expert in your industry. More important, you'll be able to ask for and get a good and fair deal for your invention.

FOLLOW THE DISTRIBUTION CHANNELS

An excellent method for exploring an industry, identifying its key people, and scoping out potential licensees is to backtrack through the various channels of distribution for a product from the end user to the manufacturer.

The following chart shows the various channels of distribution that products typically flow through. Learning the channels of distribution for your invention helps you to understand about markets and the positioning of the various companies in your industry. This information is invaluable in identifying the key people in the companies that may license or buy your invention. Right now, though, you're just researching the industry and potential markets.

Distribution channels vary from trade to trade and from account to account. The number of steps between manufacturer and retailer may vary from two (manufacturer directly to retailer) to five (manufacturer to importer to distributor to jobber to retailer). Sometimes manufacturers sell their products to distributors or wholesalers, sometimes they sell directly to major retailers, and sometimes they sell to both. Or they may sell to specific markets such as mail-order catalogs or export. We will focus on the distribution channel shown in the chart.

WORDS OF WISDOM

Be sure to date and document in writing the important aspects of your interviews with key people along the distribution channels. Chart the channels of distribution and note the size of the various markets and the manufacturers that supply those markets. You can determine the size of a particular market by multiplying the number of retail outlets in that industry by the estimated sales for similar products. This charting will help you later as you clarify your licensing strategy.

Seek out the key people at each distribution step. At retail locations, this might be a sales clerk or manager. Ask the retail manager for contact names of the store's distributors and wholesalers and their salespeople. With wholesalers and distributors, you might talk to the owner, a general manager, a buyer, or a sales manager. In the case of manufacturers, ask for the president, owner, general manager, vice president of marketing, a buyer, project manager, or the head of the new product development department—whoever is responsible for

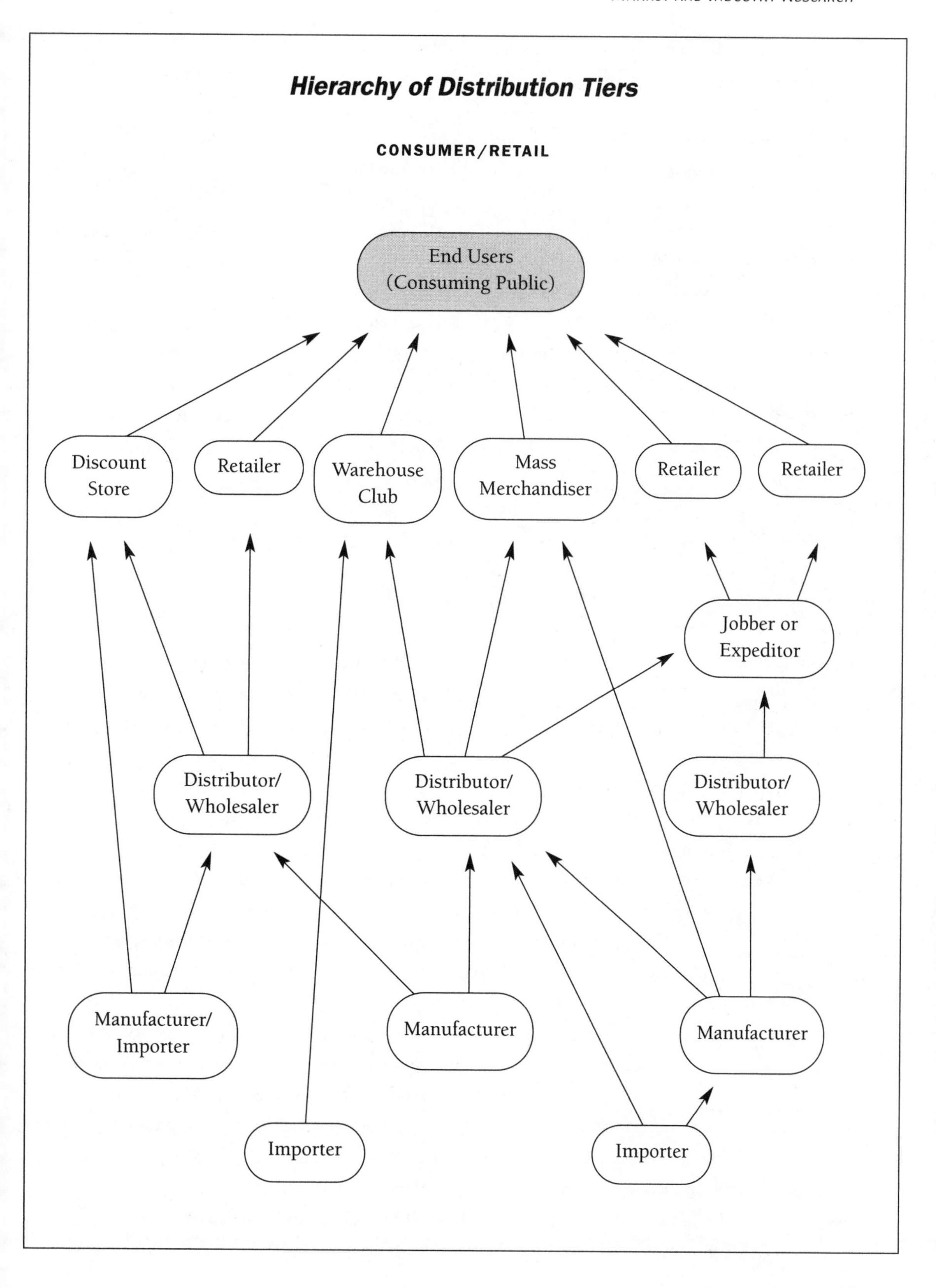
Hierarchy of Distribution Tiers
CONSUMER/RETAIL
End Users (Consuming Public)
Discount Store
Retailer
Warehouse Club
Mass Merchandiser
Retailer
Retailer
Jobber or Expeditor
Distributor/ Wholesaler
Distributor/ Wholesaler
Distributor/ Wholesaler
Manufacturer/ Importer
Manufacturer
Manufacturer
Importer
Importer

deciding which new products the manufacturer will add to its product line. Manufacturers' representatives and salespeople may also prove helpful in your quest for information about the industry.

In industrial markets, the plant manager is like the store clerk in the retail level of distribution. In the medical field, the "store clerk" is the doctor who sees the patient (the end user). Although the job titles change from industry to industry, the job functions are similar when it comes to distributing products, technology, and services. You just need to figure out who controls the levels of distribution for your particular industry.

> **KEY PEOPLE ARE THE KEY**
>
> Why are key people so important to your market research strategy? Key people on the distribution and manufacturing levels will necessarily have several years of experience in their field. You don't operate a distribution center or become vice president of a company without good experience, proven performance, and insight into your industry. Your key contact is that person who is responsible for making decisions regarding which products a retailer, wholesaler, or distributorship chooses to sell or a manufacturer chooses to produce. You or your marketing agent can leverage the expertise of these key people so you don't reinvent the wheel.

When you contact someone for the first time, it's helpful if you have a reference from someone in the distribution chain. This sets the stage for a more friendly interview. When distributors, for example, knows that one of their customers (the retail store manager) referred you to them, they are more likely to spend some time with you. Always ask your contacts if you can use their name, though. You'll almost always get permission, but this simple courtesy will keep you on good terms with all your contacts.

As you go through the various channels of distribution, your primary objective is to gather as much information as possible about the specific marketplace for your invention. A secondary objective is to start identifying potential licensees—though you won't submit your invention to them yet. Since you are just gathering information, you needn't divulge the exact nature of your invention. If you can safely divulge information about your invention (for instance, when your patent is strong), it may help you garner candid impressions of your invention from people in your field.

Start by identifying your end user.

Having people sign confidential disclosure agreements (CDAs) may not be critical at this step. Trust your gut when deciding whether or not to ask a store manager or manufacturer's representative to sign a CDA. Is it more important at this point to be safe or to get specific feedback about your invention? If your invention is truly a revolutionary multimillion-dollar item, you may lean toward protection. Most store managers and manufacturer's reps will gladly sign CDAs. But if your invention is not the next flip-top can opener, just go after general market feedback. Reveal the details of your product to only the choice trustworthy few who give you the best information. Chapter 5 has information about confidential disclosure agreements, which can help you decide what information to divulge, when, and to whom.

> **INDUSTRY SEARCH**
> If you don't know the right industry for your invention, or you want to determine all the possible applications for your invention, start by writing down all the applications you can think of and where you might use such an application. Then contact someone in that industrial area.

WHO WILL BUY?

The first step in mapping out your invention's marketplace is to consider who the end user (or consumer) will be for your invention. For consumer-oriented inventions such as a household tool or gadget, it could be homemakers, teenagers, senior citizens, or repair people. With industrial inventions, the ultimate purchaser would likely be a plant manager or a corporate buyer. List all the potential end users you can think of, and as you interview people in the industry, ask them who *they* think would use your invention. They may come up with some good ideas you missed.

WHO WILL SELL?

Next, consider where these end users might purchase your invention. Would they purchase it from a mass merchandiser (also known as a *discounter*), a department store, a hardware store, a specialty shop such as a toy store or fish store, a mail-order or online catalog, or somewhere else? Industrial buyers have their own sources, including manufacturers' catalogs and representatives, and marketing forums, such as trade publications and trade shows.

Your invention may sell in several markets. For example, your new automotive part may have applications with both original equipment manufacturers such as General Motors or Ford and in the automotive aftermarket (that is, auto parts stores). In this case, you would need to explore both channels of distribution.

Market Position

All too often inventors receive some incarnation of the following response from a company: "We think you have a very nice invention; however, it does not fit our product line at this time." What does this mean? Usually it is a very polite way of saying "thanks, but no thanks." However, such a rejection is also a way of letting the inventor know the invention does not support the company's market position.

Market positioning is an intricate part of any progressive company's marketing strategy. A company's position in the marketplace is determined by the types of products and/or services it offers, the unique features it brings to the market, and its reputation with consumers, distributors, and others in the pipeline. For example, when you think about low-price leaders in retail, you probably think of Wal-Mart and Target. These companies have established their market position well. At the other end of the spectrum are department stores such as Neiman Marcus and Saks Fifth Avenue, which have positioned themselves as suppliers of high-end, pricier merchandise. And it doesn't stop with companies. States position themselves as good bets for a variety of industries, such as tourism, and you can even position yourself in a certain way when job hunting or courting.

Successful manufacturers usually have notable market positions. For example, Rubbermaid is known as a low-price leader for rubber and plastic products made for the housewares and hardware industries. This is a well-established position for the company. It is unlikely that Rubbermaid would introduce a line of wood and metal products or high-end specialty products. Manufacturers may use the force of their market position to maintain their competitive posture or establish themselves in another industry. Therefore, the way the company is perceived by its industry is crucial.

When you present your product to a company concerned about market position, your product will have to match, enhance, and support the company's market position to get noticed. As you interview key players and do your market research (as discussed later in the book), remember to be aware of market positioning. For more detailed information about positioning, pick up a copy of *Positioning: The Battle for Your Mind* by Al Ries and Jack Trout (McGraw-Hill Trade, 2000).

It is a good morning exercise for a research scientist to discard a pet hypothesis every day before breakfast.

—KONRAD LORENZ

The Consumer Product Pipeline

Let's start our journey through the channels of distribution, using my own first invention, the Docie wedge mirror, as an example. Our first stop will be retail outlets where a consumer might purchase my invention: in this case, Kmart, Wal-Mart, NAPA Auto Parts, Western Auto, Car Quest, Pep Boys, and Auto Zone. Try to get a good mix of outlets—chains and independents, specialized and more general stores.

When you first walk in the door, spend some time looking around.

- Look at other products on the shelves where your product might be found.
- Pay attention to the amount of shelf space given to your type of product.
- Notice how many different manufacturers are supplying products similar to your invention.
- Make note of manufacturers that use materials and manufacturing processes your product will need.
- Notice which manufacturers dominate shelf space as well as those that offer only one or two items.

- Notice the features and benefits emphasized on the packaging and any collateral promotional material.
- Document the pricing of similar products.
- Write down the names and addresses of the manufacturers from the packages.

Now find an experienced store clerk or store manager to interview. At this stage in your interviewing process, it's a good idea to be honest about your intentions. Explain that you are an inventor and have a new product. It is easy (and important) to talk about the general nature of your invention without revealing exactly what it is. For example, say, "I have a blind spot mirror for automobiles that has a design feature that improves the field of view." It would be nearly impossible for anyone to guess your exact design, even when you go on to explain that it is an improved stick-on mirror for outside mirrors on vehicles that offers 30 percent greater viewing image without taking up any more space on the existing mirror. This description mentions two of the advantages of the invention without revealing design specifics. It also lets the clerk or manager know you are talking about vehicle mirrors and, more specifically, add-on accessories. By discussing a particular feature, such as greater view, you provide a basis for the clerk to comment on consumer demand for such a feature.

Clerks or managers may respond positively by stating that your idea sounds like a needed improvement and that they perceive a large call for it. On the other hand, the response may be negative or indifferent. Whatever the response, try to get them to talk about *why* they feel that way. And don't let a positive response excite you too much or a negative one stop you in your tracks. Each interviewee is one person in one store in one part of the country representing one link of the distribution chain. You have lots of other folks to talk to.

When inventors identify themselves as such, retailers, distributors, and manufacturer representatives normally respond positively and inquisitively. Inventors are enjoying a more credible image than ever before. Manufacturers that are potential licensees, however, should be approached differently, as will be explained in a later chapter.

More helpfully, interviewees may name other products that are purported to offer similar advantages. The manufacturers of these products may be either your potential competition or your potential licensees. Make note of these manufacturers. You'll want to research their product lines further and, perhaps, eventually talk to their key decision makers. Don't forget to ask to see these similar products if the store has them in stock. If it doesn't have them in stock, find out why. Are they sold out? Don't they sell well? Not right for this kind of store? This is all valuable information.

Select potential licensees that have products similar to yours; you will improve, enhance, or complement their existing product line and their market position. Pay special attention to such manufacturers as you conduct your research.

Have a list of questions ready for store clerks and managers. Be sure to ask:

- Have they ever seen products with advantages similar to those of your invention?
- What trends do they see in the marketplace with items like yours?
- Which manufacturers have a dominant presence in this market?
- Which manufacturers do they think offer the best service and sales support?
- Who sells them products of this nature, e.g., manufacturer representatives, company salespeople, distributors, wholesalers, jobbers, direct from the manufacturer, or direct import by the retailer? (Write the answers down. These will be your next interviewees.)
- What are the name and contact information of the buyer for their company?
- What amount of markup is standard for these types of products, e.g., 100 percent, 50 percent, 30 percent?
- What are the retail prices for similar products?
- To what degree does price affect sales?
- What other factors (size, material, special features) affect sales?
- How often do they reorder products of this nature, and how many do they typically order at a time? (Some products have large turnovers of five to ten times per year, while others may only turn on the shelf two or three times per year.)
- What types of packaging do the products use? How bulky is the packaging? What kinds of instructions or other copy appears on the packaging? To what degree do the differences in packaging affect sales?
- Are there particular problems that are inherent to this product area? For example, certain product areas may experience more returns than others, or the retailer may experience problems with product liability and quality control with certain types of items or with certain manufacturers.
- What mail-order catalogs and websites serve this trade? Where does one find hard-to-find items?
- What trade associations and trade publications serve this field? When are the next regional or national trade shows? Do they attend? Get names, addresses, URLs, and phone numbers.

- And last but not least: If a manufacturer were to introduce a new item along the lines of your invention, which manufacturer do they think it would likely be?

This last question is critical. The people you are interviewing will be basing their answer on years of experience in the industry. They will be considering numerous factors, including which companies have introduced products in the past along these lines and which ones have had successful track records.

Interviewing store managers and clerks (or in the case of industrial inventions, plant managers and engineers) is crucial. It gives you a real, up-to-date image of the marketplace for your invention. It also helps you practice interviewing so you can communicate effectively with others along the distribution channels. When obtaining the names, addresses, and phone numbers of distributors, salespeople, and others in the distribution channel to contact, ask permission to use them as a reference.

DISTRIBUTORS AND WHOLESALERS

As you move through the channel of distribution, the sales procedure at each step is much the same: A manufacturer's salesperson sells to the distributor's or wholesaler's buyer (or purchasing agent), the wholesaler's salesperson sells to the jobber's buyer, the jobber's salesperson sells to the retailer's buyer, and so on.

From the store level, your next step might be interviews with a buyer for that store, salespeople for wholesalers or distributors, or the manufacturer's representative, if the distribution channel is a short one.

WORDS OF WISDOM

I talk a lot about networking with your industry's key people throughout this book. Although you should seek information primarily from experts in your field who have a business and licensing perspective, there may be other experts, who are not company executives or key decision makers, who can help you. A journalist, engineer, consultant, government official, professor, or writer may have spent years in your field, observing the decision-making process. That person's insight and advice could be invaluable. Find out who the experts are and give them a call.

THE MANUFACTURER'S REPRESENTATIVE AND THE COMPANY SALESPERSON

Either a company salesperson or a manufacturer's representative calls on wholesalers and/or retailers. The distinction between these two is that a company salesperson is on staff for a manufacturer and sells only that company's products, while a manufacturer's representative represents several (six to twenty) different manufacturers that are in the same field and not competing with each other. For example, a manufacturer's representative in the automotive aftermarket may sell brake shoes, wipers, mirrors, gas caps, batteries, and wheels, none of which directly compete against each other.

Distributors usually have outside salespeople who service customers in the field and inside salespeople who service customers over the telephone. It may be helpful to know how sales staff is organized along the channels of distribution, particularly if introduction of your invention will require an extraordinary effort by the sales staff.

Larger manufacturers generally have their own salespeople so they can maintain control over their sales force. Being familiar with a manufacturer's sales force may help you decide which manufacture will best serve your invention's needs. When you interview salespeople and, later, manufacturers, ask them questions about their sales force.

Talk to whoever is selling directly for the companies you're researching. You may find that salespeople who sell for one company and those who sell for a number of companies have somewhat different perspectives on the marketplace. Always try to take into account the possible biases or blind spots of the person you're interviewing.

Many of the same questions that you asked the retailer can be used with buyers, salespeople, and manufacturer reps. However, buyers and salespeople for distributors, wholesalers, jobbers, manufacturers, and importers will have a better perspective on information such as which manufacturers have a dominant presence in the overall market. A retailer may know a great deal about selling products in their geographic location, but distributors will know more about the whole region, if not the overall market.

For instance, some manufacturers might have a strong market presence east of the Mississippi, while a completely different set of manufacturers are stronger on the West Coast. You may find that four or five manufacturers command 95 percent of the entire marketplace. On the other hand, you may find that there are twenty to thirty manufacturers serving your industry, none of which have more than 5 percent of the marketshare. You may learn that a company that seems like a good potential licensee is having problems in the marketplace because of competition from other more aggressive companies. Even though a company seems to have a good presence in the marketplace and is well perceived by retailers, it may have been sliding downward in the marketplace for several years. Another company noted for its aggressive marketing may be rapidly capturing marketshare in the industry. It is common for distributors and manufacturer representatives to have a grasp of this sort of information. Here are some additional questions to ask at these levels:

- Which companies do you have experience with?
- If this were your invention, which manufacturers would you work with and why? What are the pros and cons of dealing with those companies?
- Where are the manufacturers' greatest market presences, both geographically and marketwise, e.g., with industrial accounts, small retailers, mass marketers?

- What are the manufacturers' market positions?
- Are there rivalries between companies?
- Which companies are considered cutthroat players?
- Which companies have a better presence in which foreign markets?
- Who are the more prominent foreign manufacturers, and are they making an impact in the United States?
- What trends do you see operating in this industry, in the overall market, and for products in this category? What are the barriers?

From the answers to these questions, you may think of other and more specific questions. Place particular emphasis on recognizing the features and pricing that drive products in this industry. Compare these elements to your invention's features and benefits. Which features pique the salesperson's interest? This is what you will want to emphasize when selling your product.

Pull-Through Sales

A factor affecting marketability, and to some extent licensability, is the amount of pull-through required for your invention. Pull-through is achieved when you create consumer excitement and demand for your product so it sells. Products in catalogs and products sitting on a store shelf have to sell themselves for the most part. Only a few products have advertising budgets earmarked specifically for them. To what degree will your product or technology require that extra promotional punch or pull-through?

There are tons of ways to generate pull-through sales: through advertising, press releases, events, contests, sponsorships (as in auto racing), meetings and incentives for salespeople, and so on. When you are interviewing people in your industry, find out what degree of pull-through might be needed for an invention like yours and which companies do the most effective job at it.

So when is pull-through needed? Some products need a promotional shot in the arm targeted toward the consumer, and others need to aim at purchasing agents for mass merchandisers. This was the case with my Docie wedge mirror. I knew from our test market that consumers would purchase my invention if I could get it in front of them on the store shelf. Unfortunately store buyers didn't know this, nor did they see any demand because consumers didn't know the product existed.

Creating excitement can make all the difference.

To get the word out, I sat down with the major corporate buyers to share my enthusiasm and my test-market results before my licensees approached them to sell the product. Regional test markets in smaller chain stores showing consistently positive results primed the major chain store buyers and helped pull through my licensee's sales to the buyer. Purchases of the wedge mirror by the consumer proved to be as expected, and not a penny was spent on advertising.

WORDS OF WISDOM

Selecting an appropriate licensee comes down to understanding the needs of your project and matching them with the resources available from your target group of licensees. This understanding starts with your initial interviewing process. As you go through the process, the knowledge you gain helps you refine and emphasize those specific issues and questions most important to your project's needs.

If your invention requires more investment to pull through its sales than a licensee has budgeted, it may not sign with you. This is why inventors often start their own companies to control and provide adequate resources for pull-through efforts and successful sales.

Prequalifying the market and potential licensees is important, as is understanding how your invention will best be marketed. Although you would normally rely on your licensee to deal with this, you should understand it so you can better select companies that will effectively present and sell your product.

Mail-Order and Online Markets

Mail-order catalogs and online sales can be the optimum market outlet for a product and are often excellent secondary markets. Catalogs and websites are both forms of mail order. The main difference is how they reach the consumer with their sales message. The channels of distribution for mail-order markets are simple: the end-user purchases from the mail-order catalog company, which purchases from a manufacturer. Some small mail-order companies may purchase from distributors rather than directly from manufacturers if their sales are too low to meet the manufacturer's minimum order requirements.

In the mail-order market, the only key person standing between the end user and the manufacturer is the mail-order house's buyer, manager, or owner. At small mail-order houses, the owner or general manager is usually very accessible; in larger mail-order companies there may be one or more buyers, and each buyer will represent a specific segment of the company's product line. Your line of questioning will parallel your interviews with the retail store clerk, distributor, and salesperson. The difference is that the mail-order representative typically plays the role of both distributor and retailer.

FINDING AND INTERVIEWING MAIL-ORDER COMPANIES

There are two main things to bear in mind when interviewing mail-order catalog companies. With only a few exceptions, they don't produce the stuff; they simply buy it and resell it. And they generally do not like to reveal their manufacturing sources. Many mail-order catalog companies have a policy to not reveal the names of their manufacturers. They don't want consumers to circumvent them and purchase a product direct from the manufacturer.

The best way to circumvent this protective policy is to be honest and let the person know that you are an inventor and you are simply trying to find a company to manufacture your invention, which in turn might supply the product to the mail-order company. This will generally put your interviewees at ease, so they will share information or opinions about your invention's potential and help point you to an appropriate manufacturer. Or they might contact a manufacturer for you and pass on your contact information. Just talking with the buyers at one or two mail-order companies may expand your list of potential contacts tenfold. Although companies are reluctant to advertise their competitors, they will usually name one or two. Sometimes they will discuss the relative marketshare of the various mail-order catalogs (if they know), but they may give you an idea of the relative size of the various companies. In other words, they may not tell you that their marketshare is 20 percent; however, they may indicate that they are number three in volume for that segment of the business. Whatever you learn, you are developing an understanding of the makeup of this distribution channel.

Trade association executives and trade publications staff are other good sources for identifying mail-order catalogs specific to your market area. Specialty mail-order catalogs commonly advertise in the classified sections of relevant trade publications. The person who sells advertising space for the trade publication can be an excellent contact for tracking down mail-order catalogs, as well as discussing potential licensees and consumer preferences for product features.

Another way to locate mail-order companies is to consider the type of people who generally receive catalogs or shop online. University professors, hobbyists, and specialty retail store managers may favor certain mail-order companies. If you locate any of these people, interview them as well.

Crowdfunding, Crowdsourcing, and Open Innovation

Crowdfunding is the most revolutionary marketing phenomenon to affect inventors and consumers. Kickstarter was the first, starting in 2009, to "Bring Creative Projects to Life." There are now over seven hundred other crowdfunding websites worldwide, reported to have raised $5 billion in 2014.

Crowdfunding is the process of funding your project (or invention) by soliciting the public (crowd) via the Internet. To seek funding, you produce a multimedia presentation to place on the crowdfunding website. Your presentation describes your project and how you will use the money. If the people who view the site like your concept, they can donate to you. Donations usually start at $10 per person. In return, you offer rewards for those who donate, such as a thank-you card, or the product itself, almost like a preorder of your invention. It is not uncommon to raise from $10,000 to several million in thirty to sixty days. Popular crowdfunding websites are www.kickstarter.com, www.indiegogo.com, and for equity funding, see AngelList at www.angel.co.

Quirky, Inc. (www.quirky.com/how-it-works) incorporates "open innovation" to help commercialize inventor's projects. It accepts new product ideas from inventors. Then the public votes online as to whether they like the concept. The most popular and feasible ideas are advanced for product development, and finalists are sold to the public. The influential idea submitters receive a share of sale proceeds. Quirky sells its products in stores like Target and Best Buy, where you will sometimes see a "Quirky" section.

Crowdfunding and crowdsourcing provide inventors with a forum to test the consumer demand for their products prior to investing in manufacturing, and maybe even before patenting. Companies like Trident Design (www.trident-design.com) help inventors create multimedia presentations for crowdfunding. To learn more, see Crowdsourcing, Crowdfunding, and Open Innovation in the Resources section (page 250).

DOUBLE THE FUN

Industrial inventions, such as chemical processes or new ways of treating raw materials, often have diverse uses that span many markets. This can greatly complicate identification of potential licensees. You will follow the same general principles explained in "The Consumer Product Pipeline." The difference is that you will be dealing with key people in several different markets. Treat each market like a separate project, even though you are dealing with one process or product. If you have five distinct applications for your invention, then you have five times the work and five times the opportunity for success or failure.

Interviewing for Industrial Inventions

Inventions that fall into the industrial category are processes and products that the consumer would not normally buy directly. Examples are a process that would improve the efficiency of a printing press or a chemical composition that enhances the binding power of an adhesive. You might license to a manufacturer that would directly implement your invention at a facility, or you might license to a manufacturer that would in turn sell your product to another manufacturer.

Working your way backward through an industrial distribution chain, you'll usually start with the end user, a plant manager for example, then move to manufacturers and even raw material suppliers. The distribution chain tends to be shorter, but it can be harder to find the plant manager to start your research. If you're an industrial inventor, you probably have at least one industrial use in mind for your invention, and perhaps even some experience and contacts in your industry. If you don't know your way around the industry, trade shows and trade associations can help get you started. Some of the questions listed for consumer markets are appropriate when interviewing end users of your industrial invention. You may delve deeper into additional areas. The following questions may help guide your interviews:

- How will this technical change be embraced by engineers and other plant managers?
- Who makes the buying recommendation and who makes the buying decision?
- Do OSHA, EPA, or other standards or testing need to be met for approval for this application?
- How will this technical change affect the company's operations, positively and negatively?
- How will this technical change affect the company's profit?

ADDING VALUE

In addition to learning the market's interest in your invention, your research may disclose a benefit to the manufacturer. Implementing your invention may, in and of itself, be a minor change to a plant's process, but the effects could make a major difference to the company's overall market. For example, your improvement might enable a manufacturer to increase speed on the production line by 2 percent. In a highly competitive market, this could increase the manufacturer's marketshare by 10 percent. If you are dealing in a $100 million per year market, this would be significant. It is important to know the overall impact your invention may have on the manufacturer since they will truly reflect on the value of your invention.

Your general knowledge in your area of expertise may be as valuable as your invention itself. You might make more money in consultation fees to a manufacturer than by selling or licensing your invention. The value of your patent is limited to its claims and the know-how in its teachings. However, if you have become an expert in your field, you may possess know-how that goes beyond the scope of your patent. In the course of doing your interviewing, try to ascertain the current level of expertise in your industry. If you find that you are solving problems or even finding problems that no one else has addressed, this can be an important bargaining point to incorporate in licensing discussions. And with corporations reorganizing, downsizing, and reducing research and development budgets, you may find yourself in the expert position more and more often.

In your initial interviews, you don't have to disclose how you would solve certain problems, just make key people in your industry aware that you have solutions to these problems. This approach is critical when dealing with trade secrets and unpatented inventions. When interviewing, describe the problems you can solve and pay close attention to the responses.

If the plant managers or applications engineers are eager to know more, try to find out what they think the value of your expertise would be to your industry. Ask them for an indication as to the consultant fees that you could charge. An obvious question would be: "Which companies would appreciate having this knowledge?"

Trade Associations

Trade associations are a particularly useful resource for inventors. They usually represent manufacturers and sometimes distributors, manufacturers' representatives, and major retailers for that trade.

Some manufacturers that were founded upon one invention and are still run by the original inventor are by their very nature inventor friendly. Members of trade associations and experienced manufacturer reps are your best bet for finding those companies. Word of mouth is usually the most expeditious way to unearth this type of information.

So how do you find the trade association most appropriate for your invention? Many major libraries have directories that list most of the trade associations in the United States and throughout the world. Even so, of the numerous trade associations in your field, only two or three may be appropriate for your needs. Refer to your interviews with distributors, manufacturer representatives, and retail store managers for the names and contact information of pertinent trade associations. Examples of trade associations include the American Gas Association, the Barbecue Industry Association, the Contact Lens Manufacturers Association, the National Hardware Manufacturers Association, and so on. As you can see, associations can be broad or narrow in scope. Search for trade associations on the Internet, go to www.gale.cengage.com/DirectoryLibrary/GML33507EA%20GDL.pdf, or check the *Encyclopedia of Associations* in a library for more information.

Trade associations often list manufacturers by their areas of expertise, and some have databases with the same information. Many trade associations collect important market data on their trade. This data may include information about sales volumes; trends in the industry; breakdown in sales by type of product, by manufacturer, or by region; and more.

Trade associations also frequently publish industry directories, which contain lots of information about an industry offered nowhere else, such as company names, addresses, and toll free numbers, as well as size, types of products offered, years in business, members of the sales team from the president right down to each salesperson, sometimes even giving their districts and other pertinent information. Some directories have indexes that list companies by product category so you can quickly identify those companies pertinent to your invention. Trade associations that have websites will often list this information online.

Trade association members pay to list their company's data and to keep it updated. Information on the Internet will probably be the most current. Library directories are usually at least a year old, and government information can be up to five years old.

Other sources, such as SIC (Standard Industrial Classification—a three- to six-digit code the U.S. government assigns to all product categories—recently replaced by the North American Industry Classification System, NAICS) codes, mailing list houses, library databases, and other types of databases are less complete than trade association directories and are often outdated by at least two years.

If your trade does not offer directories specific to your needs, sometimes a phone call to the executive director of your trade association will provide valuable information. Bear in mind that the purpose of trade associations is, among other things, to help promote enterprise within their trade. When trade association executives steer inventors toward manufacturers, they are providing an important service that may help you, the manufacturer, and ultimately the trade. Many executive directors see it as part of their job to share information about the trade, including ideas about potential licensees and the names of contact people in your vicinity, such as manufacturer reps or other salespeople, who could critique your invention and give you additional leads to potential licensees.

Sometimes executive directors of a trade association only want to serve those people who are paid members of their association and therefore will not give inventors or other nontrade members the time of day, or will simply insist they purchase whatever list may be available without offering any other support. Each trade is different, and it is worth at least a phone call to the executive director; hopefully that person is in the first camp. Ask executive directors the same questions you might ask a manufacturer's representative or a distributor. Which manufacturers do they think are the best ones to work with and why?

Trade Shows: What's in Them for You?

Trade shows are an opportunity to meet people who are intimately familiar with all aspects of your industry. Exhibitors, including manufacturers, trade publishers, and associations, are there to display their wares in specially designed trade show booths, providing one-stop shopping for potential buyers.

The attendees of the show, who include buyers from major retail outlets and distributorships, manufacturer representatives, and media representatives, come to check out the goods and meet with manufacturers. Conferences and workshops provide information about the trade such as marketing and merchandising trends in the marketplace and particular problems arising in the industry. In short, trade shows can give you a quick, in-depth, and up-to-date snapshot of the state of your industry. If you haven't been to one lately, if at all

For a real eye opener, attend your industry's trade show.

> **TRADE PUBLICATIONS**
>
> Publishers of trade-specific books and magazines often exhibit at trade shows, giving you an opportunity to talk to the editors about their experience with products like yours. If you're planning to manufacture and sell your own invention, promoting it in the new product section of trade publications may be a part of your marketing strategy. If you don't meet them at the trade show, telephone trade editors and publishers in the course of your interviewing; they are usually good sources for acquiring the lowdown on potential licensees.
>
> In general, reading trade publications should be part of your research process. Their information is more extensive and up to date than information you would get at a library or through a government printing office.

possible, go! (Read on for tips on how to attend trade shows on a pauper's budget.)

CHOOSING AN APPROPRIATE TRADE SHOW

Some industries, such as the automotive parts and accessories industry, have a large, annual, national trade show. This particular show is so big (over two thousand exhibitors) that it can take two days to physically walk past every booth—more if you stop to talk.

In this industry, there are also several regional and specialty market shows. For example, one manufacturer and distributor of truck camper tops has its own independent show that brings in hundreds of dealers from throughout the world. In the eastern United States, warehouse distributors meet annually at a show geared toward products sold through their channels of distribution. This primarily covers people in the East and Midwest. This is just one industry example; different industries organize their trade shows differently.

Decide which trade shows to attend based on your budget, when and where the shows will be held, whom you'd like to meet, and what you want to learn. Contact a trade association and one or more manufacturer rep organizations in your industry to learn more about trade shows. I have found these people very willing to share trade show information with independent inventors.

GETTING THE MOST FROM TRADE SHOWS

As an inventor with a new product, there are two basic ways you can take advantage of a trade show—as an exhibitor with a booth, or as an attendee walking through the show. It is usually not appropriate to exhibit at trade shows unless your product is in production and you can accept orders at your booth. An exception to this would be when you want to test market your protected invention to the trade. This requires something of an entrepreneurial effort and, often, a sizable investment. See the section on test marketing later in this chapter.

The more prudent use for a major trade show in the beginning of invention development is simply to attend the show rather than exhibit. This can be a very effective way to collect market research information about a trade. At this stage in the game, a trade show is a place to:

SHOULD YOU EXHIBIT AT A TRADE SHOW?

Many times inventors think they need to have an exhibitor's booth at a trade show to promote their invention. But exhibitors are primarily "stable" companies with numerous products to offer. When major buyers order a product at a trade show, they expect the company to stay in business and fulfill the order. This is generally not a forum for testing the waters with your new invention. The premature introduction of your invention may dampen a potential licensee's enthusiasm that sees that your product has already been introduced to the market.

What is more appropriate is to attend a trade show, walk the aisles, and talk to the people in your industry. Learn as much as possible about your industry, product lines pertaining to your invention, and competitive factors.

If you want to get a consumer response to your invention and order commitments on a smaller scale, sometimes local inventor organizations hold shows at local malls on an annual or semiannual basis. The annual Minnesota Inventor's Congress and the Yankee Invention Exposition in Connecticut are good examples of this. These are opportunities to display your invention and potentially get exposure on national television. If you determine that entering foreign markets is important to your overall commercialization strategy, do this only if you have a patent or a patent application pending. Offering your unpatented invention for sale or publicly in this way will start the clock for the one-year time bar on your patent application and interfere with any chances for foreign patent rights.

Bear in mind, however, that you won't necessarily meet the who's who in your industry or learn the shakedown of that industry at a trade show geared toward inventors. Consider an inventor's show as a dry run for a major trade show and an opportunity to do some test marketing.

- Gather valuable information about the marketplace as it pertains to your invention and about trends in the marketplace.
- Get a critique of your invention and valuable feedback from key people at the various booths (if you're far enough along in the proprietary protection process).
- Identify and have conversations with manufacturers that are potential licensees for your invention.

Remember that exhibitors' primary purpose at a trade show is to sell their products. It is not uncommon for exhibitors and trade show officials to take offense when an attendee appears to be pitching a product. Be considerate of other people's time and emphasize the market research nature of your motives.

When walking the aisles of a trade show, pay attention to those manufacturers that offer items that would be complementary to your invention. Take note of the size and sophistication of those manufacturers.

CONTACTING POTENTIAL LICENSEES AT TRADE SHOWS

Your first contact with a potential licensee may come at a trade show. Since you are still gathering information and not yet pitching your new product idea, it is not critical that you speak with the key decision maker. However, if the opportunity presents itself, an initial conversation with the key decision maker can be extremely valuable. Not only will you learn more about that person's company, but you can also present your invention, and yourself, to a possible licensee. *Note:* Before you approach a potential licensee at a trade show, be sure you've read Chapter 5 of this book, on preparing to submit to potential licensees. This chapter explains what you should try to learn from an initial contact and how much is safe to reveal.

Trade show booths can be deceiving. Although you can usually judge the size and scope of a company by the product line displayed in its booth, sometimes companies that are dominant in the marketplace only set up small booths to emphasize a few of their newer items rather than their whole line of products. So of the five thousand products the company offers, it may only display the fifty newest products at the show. Conversely, smaller companies that want to evoke a competitive image may display in a very large booth. Watch out for this head fake.

A good opening line when you approach someone at a booth is to introduce yourself and then ask: "Who in your company is responsible for deciding which new products will be introduced into the product line?" Then ask whether these people are at the trade show, and if so whether they are available to speak to you. If they are not at the show, try to find out where they can be reached. Call them from the trade show to tell them that you want to introduce a new product to their company. Ask them to refer you to someone at the booth who can look at or discuss your new product concept.

Some of the people at the booths may spend time with you and offer a considerable amount of information, while others may be brief. In any event, it doesn't hurt to ask to speak to the key people, even if they're upper-level executives. It never ceases to amaze me. One president will blow you off, and the next one will spend thirty minutes with you. But remember, their goal at the trade show is to make sales; they can't talk to you when a customer walks up.

When you talk to someone in the booth, try to get answers to the following questions:

- Is the company interested in new product ideas in your area of invention?
- Does the person you are speaking with make the final decision about new product introduction, and are any others involved in making this decision?
- What are examples of new products that the company has introduced in the marketplace recently?

- What kind of success has the company had with its more notable new product introductions?
- What is the company's experience in dealing with outside inventors: how many, who, when, and has up-front money ever been paid out to the inventor?
- What is an example of the various royalty rates? (The objective here is to get a range, not to nail the company down to a specific figure.)
- Would the company be interested in looking at your invention?

If the answer to the last question is affirmative, you face a major strategic decision as to whether you should actually present your invention to the decision maker at the trade show or would rather discuss the formal procedure for submission to the company. You won't necessarily know the answer in advance; just be prepared for either possibility.

If it turns out that you need to go through a more formal submission procedure or you feel it's inappropriate to disclose your invention now, that's fine. Just talk about the specific advantages and benefits of your invention without giving away trade secrets and other working knowledge, and ask whether these specifics would be valuable to the company. If you are striking a positive chord, the company may be lenient about submission procedures in order to have a look at your invention.

Opportunity May Only Strike Once—If a potential licensee expresses interest in reviewing your invention, either at the show or later, try to have that person sign a confidential disclosure agreement on the spot. This sort of agreement may offer you broad protection and can strengthen your position when you submit to that company later on. They may ask you to sign one of the company's disclosure agreements before proceeding with the disclosure of your invention. Read about both kinds of disclosure agreements in Chapter 4.

On the other hand, if the key decision maker is interested in looking at your invention right away, either by signing your confidential disclosure agreement, looking at your patent or other nonconfidential information, or being disclosed confidential information in the presence of your own witness, then the next step may be to go ahead and disclose your invention. Learn more about disclosing an invention in Chapter 5. Also pay particular attention to the "Smooth Success" section in Chapter 4 for tips on this issue.

Pay close attention to initial reactions.

After you have disclosed your invention to the company's decision maker, pay close attention to that person's response. Don't pitch your invention—just watch and listen. There is an old saying in sales that at this point, the next person to speak loses. The invention needs to stand on its own merits, with the exception of any clarifying explanations that are absolutely required. Once you've given the person a chance to look over your invention and offer an initial response, the conversation may go one of two directions:

- With a positive response, you can ask what the submission process to the company would involve from this point on. Also ask what the person likes most about your invention.
- With a negative response, you can ask for a referral to another company that may be interested and in a better position to commercialize your invention. Also ask what it is about the invention that the person sees as needing improvement.

Whatever the response, you will have gained another informed perspective on your invention, and possibly a good contact with a potential licensee.

WORDS OF WISDOM

If you talk to potential licensees at a trade show and offer your invention for sale, this could set the one-year time limit (time bar) ticking with respect to obtaining a patent. In other words, you have one year from the time that you offer your invention for sale, or disclose it publicly, to file your patent application if you have not done so already.

There is a distinction between offering your invention for sale and seeking a potential licensee who would in turn commercialize your invention. However, this differentiation is a gray area; to be safe, it is important to be very clear that you are not offering your invention for sale, but are merely researching licensing possibilities.

HOW LONG WILL IT TAKE?

Allow at least one to two days to completely walk through a national trade show and seek out as much information as possible. I prefer to arrive on the second day of the trade show to give the exhibitors a chance to observe the current conditions of the market, which I hope they will then share with me during the interview. Plan to complete your business within one and a half to two days. In the case of exceptionally large trade shows, you may need to spend three or four days.

One school of thought suggests you should walk the entire trade show and get the lay of the land and then go back to those booths that pique your interest. Another method is to stop at appropriate booths on your first pass. Whatever you do, remember to wear a good pair of walking shoes. By the end of the trade show, you may have traveled five to twenty miles on foot.

WHAT WILL IT COST?

If you don't exhibit, your cost and preparation for attending a trade show, whether it be a regional or national show, should be minimal. The cost is primarily dictated by the mode of transportation you choose and how far the trade show is from your locale. Many national trade associations charge a nominal fee of $20 to $100 to attend a show. Most shows offer a discount or a waiver for preregistration, which normally has a deadline forty-five to sixty days prior to the actual show. Trade show organizers may insist that you present a business card to enter the show. They may even insist that you bring an invoice or purchase order sheet to demonstrate an intention to actually purchase goods and write orders. Usually, however, a business card is sufficient. Spend $10 or $15 at your local quick printer and establish whatever company name you choose, if you haven't done so already.

If you exhibit at a trade show, preparation will take at least three to six months. Shows usually run for at least eight hours each day, and you will need more than one person to run your booth to give each other time for meals, bathroom breaks, and the opportunity to visit other exhibits at the show. The cheapest booth at a regional show may be in the $500 range for a 10 by 10-foot booth. At major national shows, a 10 by 10-foot booth costs anywhere from $2,000 to $4,000 for a two- to five-day period. Tables, curtains, signage, electricity, telephone, waste disposal, and

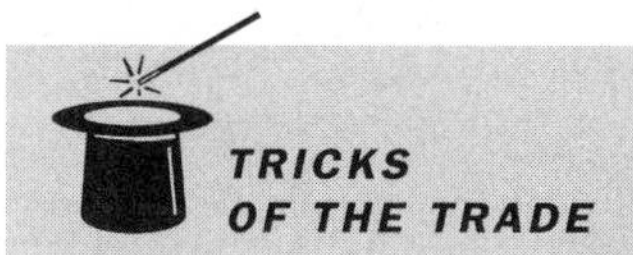

The $100 Trade Show—In the past thirty-plus years, I have perfected a method for attending several days of a major national trade show for a budget as low as $100, including transportation. Here's my modus operandi for a shoe-string trade show:

- Drive to the show. Trade shows in a five hundred–mile radius from my home are within striking distance. To keep to this budget, leave the house early on the day of the show (this eliminates the need for an overnight stay).
- Find the major companies doing big promotions and seek out their hospitality suites after the show closes for the day. This is a good way to get a free dinner. It helps to have registered as a buyer; they are the ones being invited to dinners and parties after hours. The disadvantage of being labeled a buyer is that everyone is always trying to sell to you, and it can be harder to do the work you came for!
- Stay at a hostel or YMCA for cheap, basic accommodation and a chance to talk to the world travelers who are your fellow guests.
- Use local mass transit to avoid parking fees.
- Bring snacks to avoid the inflated food prices at a convention center.

I've actually done this several times, traveling from Ohio to Chicago. It's a great value and an excellent way to spend $100. The things you learn at a major trade show and the acquaintances and friendships you make there are priceless.

all other amenities for your booth cost extra and are not cheap; the sometimes mandatory union services to install these items are generally extra. Displays, photographs, signage, printed brochures, samples, and any other promotional material that would stand or hang in your booth cost money, as do packing and shipping this material to and from the show. Making follow-up contacts with people you meet at the show is a further expense. By the time all of these costs are added together, it is hard to exhibit at any show for less than $3,000. A major national show will likely cost a minimum of $7,000 to $10,000. A large corporation's trade-show budget can go over a million dollars for one show!

Test Marketing

Test marketing is a means of testing the market's response to your product to determine sales potential and how you might most effectively achieve those sales. Many inventors come to me after having performed their own test marketing. Typically, they have gone to their friends, relatives, and neighbors and proudly shown off their invention. They ask, "Would you buy this?" Then they ask, "How much would you pay for this?" And that's it—market tested.

I did a market test like this once myself. Did I say once? This is probably the least conclusive test marketing you can do. First of all, once you have proudly proclaimed that you are the inventor of said product, your friends and relatives are not likely to tell you "your baby is ugly." So, as you might expect, 95 percent of the responses to these types of test markets are indiscriminately favorable.

It gets interesting when you discuss pricing. The simple fact is that the price people say they will pay for an item is invariably much higher than what they will actually pay. Of the people I asked about my Docie wedge mirror, 90 percent said they would pay an average of five dollars for one. In reality, we discovered that people do not want to pay more than two dollars for the passenger car version of my invention.

Many inventors fall into a trap when they get positive feedback from sampling consumers on their invention. When four out of five friends say they would buy the invention, the inventors expect that four out of five people will actually do so. However, this is rarely the case.

To accurately test market response to your product you need a substantial budget to make production models of your invention and package it appropriately. This is normally beyond the budget and expertise of an independent inventor.

Hiring a professional market research company to do formal test marketing ranges from $10,000 to $50,000 or more, depending on what aspect of your invention you are testing. For example, if you are trying to decide what type of packaging (color, shape, size, verbiage, and other such elements) will enhance the sale of your product, you may be able to build prototypes of several package designs and then display these to people who match your target markets demographically and psychographically. Demographics characterizes a population using distinctions of age, sex, nationality, and so on. For my Docie wedge mirror, the target demographic mix was primarily a male over seventeen years old who owned a vehicle with outside mirrors. Psychographics describe the population based on a person's attitude, behavior, buying patterns, and interests. If you are considering an auto accessory like I was, you might go to an auto parts store, where you will find similarities among prospective customers who currently purchase stick-on mirrors.

A professional market research firm may select a group of men over seventeen years old who own vehicles, who purchased mirrors, and who are willing to look at packaging and be paid $20 to $50 each for their trouble. This group of people is called a focus group. The focus group would consider the product's packaging and choose the one it likes best. The research firm may go a step further and ask several questions about why certain packages were preferred over others. Since packaging is such an important element in consumer sales, this is a common theme for focus groups. And you can gather information without revealing your specific product to the focus group.

Test marketing can be very effective—but only if done properly.

However, if you are doing nothing more than systematically asking a whole bunch of people about their preferences for different designs and configurations of your invention, irrespective of the packaging, then this information can prove quite helpful. If they like your product then they have little reason not to like it in an attractive package. So don't skimp on packaging.

Packaging aside, will the product sell? The most accurate way to test a product is to have it on a store shelf and see how it moves. Large chains and mass merchandisers, such as Wal-Mart, rely on sales figures from smaller stores in regional test markets before putting a product on their shelves worldwide.

As an independent inventor, you obviously don't have access to a group of ten or fifteen chain stores, but you can set up a similar test market, like I did for the Docie wedge mirror. I approached several independent automotive stores whose owners were willing to make a buying decision on the spot. We placed products on the shelves and monitored sales. We even offered incentives to store managers for questioning people when they purchased the mirror, asking why the product appealed to them. We tested the pricing by putting different prices on the mirror in different stores at different times and then compared sales figures. Sometimes you have to be a bit of a statistician to figure out how to do these tests. It's easy to misread the resulting information or get completely erroneous information.

If the notion of doing a test market is out of your comfort zone, you'll either need to hire a professional or find a licensee who will do it for you. Proper test marketing is usually done only by the most savvy companies.

So how do you get an accurate sense of sales projections for your invention? Through the years I have learned that you do not have to put your product in front of the buying public to get a ballpark idea of its potential. Key company executives, particularly those in the marketing department, have so much experience that they can just look at certain items and estimate potential sales velocity (how fast the product will disappear from the shelves). Granted, sometimes they are wrong, but key executives are a good place to start, especially if you can get their information for free (see Chapter 5 for more information about how to do this).

If it weren't for Philo T. Farnsworth, inventor of the television, we'd still be eating frozen radio dinners.
—Johnny Carson

I learned through interviewing that approximately three million stick-on, blindspot mirrors are sold annually in the United States. Once I knew that, it was simply a matter of determining what portion of that market my mirror could command. If my test market showed that consumers chose my mirror 33 percent of the time, then I could potentially sell one million units per year. Even without the test market, I knew that the potential sales were somewhere between zero and three million units per year.

Once you have estimated sales volume for your product or technology, you can figure out how your invention will fit into the market. Although test marketing doesn't give you exact data, it can help you make strategic decisions regarding how much you want to spend, or risk, in pursuing your invention.

Obviously, if you discover that the maximum potential royalties from your invention are only $1,000 per year, you might think twice about rushing out to file a patent application for $5,000, particularly when you factor in all the other expenses and any potential market risks.

The key here is to use common sense. What kind of information do you want to acquire, and who is the most appropriate group of people to acquire it from? When I did informal test marketing, I simply kept asking the same questions until I found a definitive pattern in the responses. If, in your first hundred responses, 70 percent of the people liked item A, and in the second hundred responses, 70 percent liked item A, and in the third hundred responses, 70 percent liked item A—then take a hint.

When I was working to relicense my Docie wedge mirror, I conducted a test market in my hometown. This turned out to be extremely important to my licensing activities. I designed a display board that mounted on my photography tripod. On it I displayed several wedge mirrors, some of which I had made from scratch. They were of different sizes, shapes, and materials. The purpose of the display was to determine customer preference in only these aspects.

The display was set up in prominent sections of both a NAPA and an Autoworks store on two different Saturday mornings during their busiest hours. Response was fielded from over two hundred people who had either owned or used a stick-on spot mirror. This was a successful test because the audience clearly matched my target market. I also compared consumer preference of the existing mirrors on the market to my mirror design. To my pleasant surprise, my mirror designs were preferred 78 percent over the mirrors currently on the market. Also, as a result of the test, I gathered information necessary to complete the designs of two new mirror sizes and shapes and created a whole new image for the base of my mirror. This allowed me to eventually triple my royalty income.

Key people have information that is difficult to get anywhere else.

Prequalifying Inventions—Interviewing key people in your industry can be quite valuable when you are attempting to decide which of several inventions to pursue. It enables you to ascertain whether your new idea is even worth pursuing based on the market factors that you discover. In the course of acquiring information by interviewing key people in your industry, you may find there are verifiable reasons why you should not pursue an invention any further. For example, you might discover:

- Key elements of your invention violate OSHA, EPA, or other laws.
- Other existing technologies may be superior to your own.
- The potentially patentable features of your invention will not protect it to the degree necessary to have commercial worth (this would apply particularly in highly competitive markets where the manufacturers have a track record for designing around patented inventions).
- The anticipated cost to produce your invention would be much higher than what the market will bear.

Unfortunately the list can go on and on.

When you encounter this type of information, it is important to weigh the risks of spending money to apply for a patent versus the risks involved if you don't. Don't rule out that market conditions change, manufacturing processes change, and there could be a host of other considerations that would cause your invention to have worth in the future. The decision you have to make is: "Is it worth the cost of patenting and paying the maintenance fees to wait out potential changes that might affect my invention's commercial viability?"

Why Interviewing Works

Interviewing key people in your industry provides insights you can't find in a book or a database. And it continues to build on your base of information when deciding whether to sell your invention, form your own company, or consider exclusive or non-exclusive licenses for your invention (and which manufacturers you should contact first).

What is great about leveraging the expertise of key executives is that it's so darn cheap. Not only is a big-time market research firm expensive, it is not likely to make extra calls to key executives to pick their brains. As you may imagine, this does not go over too well since company executives are busy and don't appreciate calls from a market researcher. Inventors, on the other hand, have a special angle—a carrot, if you will. The company you are interviewing may someday license, buy, sell, or otherwise make money from your invention. You may want to keep your carrot in view, especially when interviewing manufacturers for the purpose of obtaining preliminary market information. We inventors need every edge we can get!

Anticipated Costs and Timetable

One of the great advantages of using the methods described in this book to identify potential licensees is that you can acquire a vast amount of information on a modest budget. The resources necessary to gather information can be as minimal as the use of a telephone and visits to companies and stores in your closest major population area. Major metropolitan areas have retail stores and distributorships that are representative of most industries. Although you may find retail stores in close geographical proximity to you, major distributorships may be a bit farther away.

There is real value in going in person to retail outlets or to visit plant managers. In most cases, interviews with distributors and with manufacturers' representatives can be done over the telephone. When possible, however, the greatest value is achieved when you can meet with key industry people in person. Since extensive travel may be required to do this, it is important to identify the best distributors to contact while doing your retail interviews. This way you do not waste time and money traveling around to visit with people who can't or won't help you.

The timetable for acquiring this preliminary market information can be as short as a couple of weeks or as long as a few months depending on your availability and the nature of your invention.

The cost involved ranges from under $100 upward depending on the extent of your travels. However, it is important to not limit your interviewing to any one geographic area. The market conditions on the East Coast may be entirely different from the market conditions on the West Coast, for example. The telephone is a very effective way to cut time and expense when interviewing over a wide geographic area. Many distributorships and manufacturers have toll-free numbers.

A few professional companies will obtain market information in your industry for you. The charge for this type of field research typically ranges between $1,000 and $10,000 depending on the complexity of your project. A typical time frame for this service would be from two to nine months. Such companies are discussed in Chapter 7.

What's Next?

When you've finished this stage of intensive market and industry research, including interviewing key people throughout the supply chain and possibly making first contact with some manufacturers at trade shows, you should be able to answer the following questions. You are ready to move on to the next phase in the commercialization process: identifying your short list of potential licensees and picking your top choice.

THINGS TO REMEMBER AND CONSIDER

- What are your invention's possible applications?
- What are the estimated sizes of the various potential markets for each of these applications?
- Which of your invention's features are most desired by the industry?
- In what form might your product be presented to the consumer?
- What is the route your product would take through the distribution chain from manufacturing to retail?
- How much pull-through marketing will be required, and which companies can do the best job?
- Are there several market opportunities, like mail-order, online sales, etc.?
- Are there any barriers to commercialization or reasons to make a design or production change?
- Which companies have strength with which markets?
- Which companies have strong international distribution and sales in foreign markets?
- Which companies and products may be potential competitors?
- Where and when are the next trade shows scheduled?
- Whom can you next ask for help?

CASE STUDY

An Automotive Accessory (Part 3)

Back to the Docie wedge mirror saga . . . Thus far, my original patent had eventually been rejected, Rubber Queen had scrapped my product, and other manufacturers were moving in with their own wedge mirrors. Then I established a new patent position, parted ways with my first licensee, and now had sixty days to exercise a license option with my former partner/investor. My next objective was to attend the upcoming automotive parts trade show and see whether I could influence companies to license my invention. This would be especially interesting considering that the companies didn't know about the new improvement patent.

This is where preliminary market research proved helpful. From my interviewing, I already knew some of the key people in my industry and knew where to look for others. I had an idea who the major players were, so I knew whom to negotiate hard with and whose market positions were weaker.

At the APAA (Automotive Parts and Accessories Association) trade show, about eighteen different manufacturers were offering my wedge mirror for sale to auto parts stores and mass merchandisers. At each booth, I spoke with key decision makers and informed them that I had rights (or an option for rights to the new patent in this case) to a U.S. patent that their mirror might infringe on. I didn't go into specifics about what my rights were, but I showed them a copy of the patent, sans owner's name.

I basically asked each company to either license the mirror from me on a 5 percent royalty basis or discontinue the product. Two of the companies took the product off the shelf right then and there and said they were not interested in a license. They claimed that it wasn't doing enough volume for them to mess with it. One company physically escorted me out of its booth and told me that I had a lot of nerve. It never did license my mirror, but its volume was insignificant anyhow. Most of the companies were somewhat surprised by the whole phenomenon and said they needed more time and information before they could make a decision. They requested copies of the patent and a proposal. They were not interested in spending too much time at the trade show discussing this. The trade show is their time to sell to clients, not for negotiating deals.

One company that did a very high volume indicated it would be willing to consider a license—if the other major competitors were also paying royalties. This company did not want to be faced with its competitors having an unfair 5 percent price advantage; this would be enough to knock it out of its major accounts.

Another manufacturer's president was escorting me from his booth when he said, "Kid, you are either very naive or very crafty to come into my booth and submit such a proposal." In my best naive tone, I told him I was simply trying to see the product line improved and get my just due. We ended up back in his booth discussing the details of why I would be asking for a royalty on a product that his company was already offering royalty free. When he realized that I wasn't a con artist and might actually have certain patent rights, he saw an opportunity to capitalize on this and potentially be a mass distributor for the product, thereby knocking out much of his competition.

What had begun as a near kick in the pants turned out to be the leverage I needed to restart the licensing process. Now I had two companies that were clearly interested. This company, Company A, was a very large company with moderate sales of my wedge mirror. The first company, Company B, was a smaller company, but an account with Kmart gave it the greatest distribution in the marketplace. However, Company B did not want to license unless it knew that other major companies, such as Company A, also had licenses and paid equivalent royalties. Company B also wanted assurances that the other major competitors would follow suit. So it all rested on Company A to start the ball rolling.

(To be continued in Chapter 6. . .)

HINDSIGHT LESSONS

When I started talking to potential licensees at the trade show, I was civil, up-front, and showed that though I meant business, I could work with them to our mutual benefit. Establishing and maintaining good relationships with my licensees, my patent attorney, and others whom I rely on is one of the most important elements in this equation. I almost never limit my conversations with these people to business only. I can tell you a little bit about the families of the company presidents whom I deal with, their secretaries' names, how many children they have, favorite hobbies, and other personal tidbits. Most people like to make business as pleasant as possible. Stuffed shirts don't get very far. I am not suggesting that you should necessarily become buddies with the people you do business with, but you can have a little fun with them. And it is harder for people to say no to you when they like you. Conversely, it's easier for them to say no if they don't.

An inventor is simply a person who doesn't take his education too seriously. He tries and fails maybe a 1,000 times. If he succeeds once then he's in.

—CHARLES KETTERING

CHAPTER FOUR

TIME TO PICK THE LINEUP

Qualifying and Contacting Appropriate Licensees

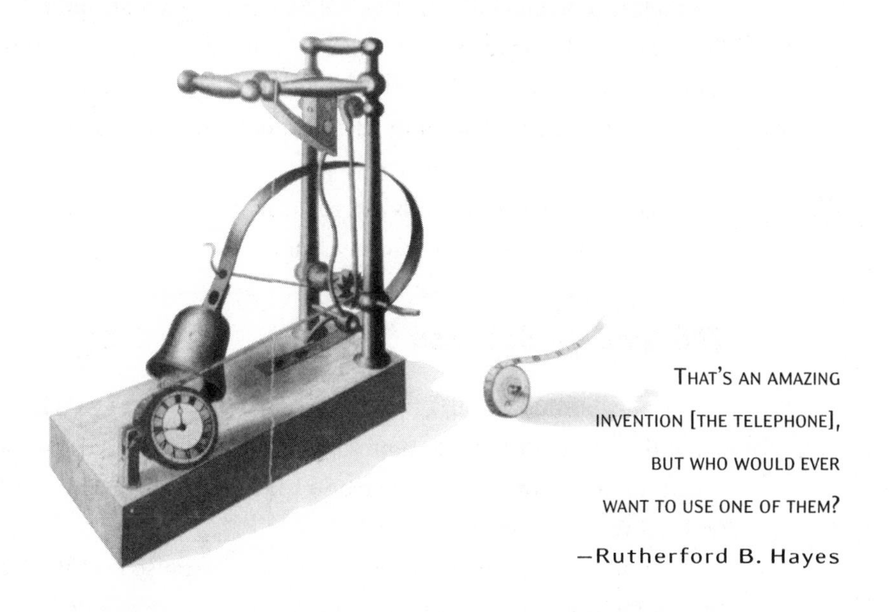

THAT'S AN AMAZING INVENTION [THE TELEPHONE], BUT WHO WOULD EVER WANT TO USE ONE OF THEM?

—Rutherford B. Hayes

In the past three chapters, you learned about the benefits of licensing your invention, considered your strategy for seeking patent protection, and did lots of market research in the industry or industries where your invention has application. You have a good idea of who the manufacturers are in your market and have conquered the previously unfathomable maze of distribution channels. In fact, you may know more about some aspects of your industry than any one of the key people you spoke to, since you have just collected the most up-to-date information from a variety of sources.

Still, you do not know how potential licensees will react to your invention or what kind of offers you will get from them. If industry response is not to your liking, you may yet choose to start your own company and test the waters.

Whether you eventually license or strike out on your own, the procedures described in the following two chapters will provide you with valuable competitive information that may not otherwise be available from other sources.

Now you are at a crucial decision point. Which manufacturing companies do you want to contact, and in what order? You do not want to arbitrarily disclose your

Alarm bell to be fixed to a clock or watch by Benjamin E. Freymuth (U.S. Patent No. 2244X), Philadelphia, Pennsylvania, 1814.

invention, patented or not, to any more companies than necessary to get the information you need to make a knowledgeable decision (since you may still be considering starting your own company). In some industries, companies are very protective about their product development efforts. These manufacturers will be more enthusiastic about promoting your invention if they know they are the first and only company you are approaching.

Another important factor to consider is that your invention may have limited proprietary protection. In other words, you may be relying on trade secrets and know-how without patent protection, or your patent may have limiting claims that some companies may find easy to design around. Plus, if you end up manufacturing on your own, you don't want to release information to a potential competitor that could be used against you later. As a practical matter, your time is valuable. These are all good reasons to carefully target the first company(s) that you make contact with and limit disclosure of your invention.

The Smooth Success

Before I get into the details of finding appropriate licensees, I'd like to share a story about the type of deal that invention developers like me live for. For the most part, everything about this deal went smoothly and was timely. This is, of course, very rare and very welcomed.

I cannot say exactly what the invention is because I want to reveal some of the behind-the-scenes, nitty-gritty dealings. Such knowledge could be used against the licensee since the product is now on the market. I can say that this is a sports invention.

When the inventor first approached me, I initially turned him down. His concept was very simple, and I doubted that any patent protection would be significant enough to be valuable. It seemed like the type of item anyone could easily replicate, and the industry was one where competitors commonly copied others.

After some arm twisting from the inventor, I accepted the project. I put it through the normal paces. First I performed market research in the field and interviewed personnel at four retail outlets. I found that there was a gap in the market for a product of this nature and that all the existing products performing the same function were priced ten to twenty times what the inventor's product could be sold for. (The inventor had supplied me with some basic cost figures for his invention to help me determine a price.)

Next I contacted two distributors. From the information I gathered, I had a good handle on the names and addresses of six potential manufacturers. I got brochures from these manufacturers and learned that none were publicly held, so detailed corporate information was sketchy at best.

I obtained confidential disclosure agreements (CDAs) from two of the manufacturers and submitted the invention to them. What I realized in my telephone follow-up with these manufacturers was that they were not the best targets for this invention; however, they both referred me to two additional companies that they thought would be good licensing candidates.

One of these companies commanded the greatest marketshare. It sold to distributors, major retail chain stores, mail-order catalogs, and the major foreign markets where this product is used. At first this company (which I'll call Company Y) appeared to be small. It turns out it was owned by a much larger company that had all the resources needed for this project.

Company Y's marketing manager signed my CDA, and I submitted the invention to them. In about three weeks the president of the company phoned me to inquire about the technology. This was a very good sign.

I explained how the invention worked and discussed its availability for license. The president's primary concern was whether or not I had submitted the invention to Company Z, which was the number two company in the industry. I explained that Company Z had signed a confidentiality agreement; however, I had not submitted the invention to them yet. The president's response was that if I submitted it to Company Z, he would no longer have any interest. He considered Company Z to be Company Y's only substantial competition, and he further stated that Company Z would not hesitate to steal or otherwise plagiarize the invention. The president was interested in having an exclusive opportunity to introduce the product to the market and get a head start over any potential competition. This meant more to him than potential patent rights. Furthermore, he wanted to use this item to showcase his company as an innovative producer of these types of products. He couldn't do this if a competitor was lurking in the wings. Even if Company Z did eventually introduce a competitive product, it would be a Johnny-come-lately entry.

Another feature the president found particularly appealing was the name the inventor had conceived for his product. When I explained that we might be able to acquire a national registered trademark for the name, this extra level of proprietary protection pleased the president. I offered to draft a licensing agreement for Company Y to review. Meanwhile, Company Y would sharpen its pencils and establish a level of minimum sales performance in exchange for an exclusive arrangement. Company Y's attorney accepted our licensing proposal, and the inventor and I met at Company Y's headquarters to ink the deal.

The entire process, from the time the inventor approached me until there was a signed licensing deal, only took about six months. It is the smoothest licensing deal I have ever orchestrated. The company was not willing to pay cash up front because it wanted to use all its resources to promote the product. It did, though, offer the inventor a substantial advance against future royalties. Either end result worked equally well for the inventor, but was an important distinction for the company to make. Semantics can make a difference. The inventor used a portion of these funds to apply for a patent and a trademark. The company immediately began the preproduction process and introduced the invention to the market at the next trade show—even before units were available for shipping.

This licensing effort was so successful because the licensing agent (in this case, yours truly) was optimistic about the chances of commercialization even though the possibility of establishing patent rights was nil. Also, the inventor accepted reasonable terms and didn't inflate the value of his invention. Approaching the number one company in this market first also helped. Had I started with the number two company, the outcome may have been less favorable, if not a total loss.

Be selective about whom you submit your invention to.

By having the patent pending during Company Y's market introduction (and possibly beyond), Company Z will not know how to design around the invention because it doesn't know the claims of the patent to design around. As of this writing, the patent is still pending, and obviously the inventor will take every strategic opportunity to keep the patent pending for as long as possible.

This is a good example of why it is crucial to do your homework and perform market research without revealing any more information than is necessary to develop the best commercialization strategy for your invention. Had I waited for the patent process to be completed, we may have had a patent rejection, and I may have lost the window of opportunity to introduce this product on a timely basis. Needless to say, we were all very pleased with the outcome.

Picking the Lineup

The "Smooth Success" story is a perfect example of why it's important to research potential manufacturers before submitting your invention. Had I gone in cold and uninformed, the outcome wouldn't have been so favorable.

Decide which company or companies to contact first based on the results of your research. You want to know as much as possible about the following aspects of each manufacturer's business.

- What is the extent of its manufacturing resources and expertise? What is its volume of production in your field? What are the size and makeup of its product development support staff, and are engineering talent and innovation a priority with them?
- What is the company's financial position? Is it well capitalized and profitable, or overextended and operating on a slim profit margin? Does it seem to be well managed?

- What is the company's market position? Are the company's philosophy and direction a good match for your invention?
- Does the company often introduce new products, and if so does it succeed with those introductions? Are new products well supported by strong sales and marketing efforts?
- Does the company have pending litigation that could adversely affect its business? Does it have any history of litigating against independent inventors?
- Is the company's presence in your targeted marketplace adequate? In other words, if it took on your invention as a product, could it move enough units to net you some decent royalties?
- What do people in the distribution chain have to say about the company? Is the overall verdict from retailers, wholesalers, distributors, and manufacturer's reps that it is a strong, well-run company with superior, well-supported products?
- What is the company's philosophy? What values are stated in the company's mission statement or implied in the ways it does business?
- As far as you and your sources can tell, is the company stable? Or is it on the verge of a management or labor shakeup, a merger, or a hostile takeover?

More Free Information—When you are dealing with large companies that are publicly held, you can call the company and request its latest annual report, including the most recent quarterly reports, its Securities and Exchange Commission (SEC) 10-K (annual) and 10-Q (quarterly) reports, and any company brochures or catalogs.

By law, publicly held companies are required to provide an SEC 10-K or 10-Q report to the public upon request. This report includes information about the sources of the company's raw materials, the location and size of its manufacturing plants, the number of employees, a complete financial breakdown including salaries, its standing in the marketplace, its market standing relative to its competition's, any pending or recent litigation or lawsuits, and any other information that would affect the company's financial health. This information is meant for people interested in purchasing stock in the company so they can make informed decisions about the risks involved. It's also very helpful as you evaluate potential licensees. And it's free. Corporate headquarters normally disseminate such information, or go to www.sec.gov.

Privately held companies may not give you an annual report, and they are not required to submit an SEC 10-K report. You may only be able to get a brochure. This is why it's so important to interview key people; they may be your only source of information about such a company. Ask them questions about the information normally contained in an SEC 10-K (or 10-Q) report.

If you have already determined all this information from your interviews and research, then you're ready to move on. If not, some more preliminary research is in order. The manufacturers themselves may be the best and only sources for much of this information.

Thus far, your primary purpose for communicating with people in your field is to obtain market research. When you start interviewing manufacturers, your primary purpose is still market research. The information in this chapter will help you further understand the manufacturers and their markets, attitudes, and interests. But don't pitch your invention quite yet! You don't want to fire without first aiming, and in this chapter we're refining our aim.

For the sake of example, let's say you have identified six different companies that seem to be appropriate commercialization candidates for your invention. They manufacture similar types of products, are in the right market, and appear to have the resources to introduce your invention. We'll also assume that you did not discuss details of any potential licensing arrangement at a trade show. We are starting from scratch with your potential licensee.

If you have found one company that commands 80 to 90 percent of the appropriate marketshare, you may want to approach it first.

On the other hand, you may feel that none of the six companies stands out more than another in the marketplace. You can simply select the one that feels best to you and start from there.

MARKETSHARE VERSUS DISTRIBUTION SHARE

It's important to recognize the difference between marketshare and distribution share. If $100 million worth of products were sold in your invention's market and Company A's sales are $80 million, then Company A commands 80 percent of the marketshare. Company B, with $10 million in sales in that industry, has only 10 percent marketshare. Company A would seem to be the perfect licensee for your invention.

However, consider this: Company A has 80 percent of the marketshare, and it has products placed in Kmart, Wal-Mart, and all the major retail outlets in the United States—thirty thousand outlets total. If Company B placed its products in nearly all of those same thirty thousand retail outlets, its distribution share would be nearly equal to that of Company A. Since your overall objective is to place your invention in as many retail outlets as possible to maximize distribution and sales, Company B may support you as well as Company A. In other words, Company A may have eight times the shelf space, but Company B would still have a presence in most of the same stores. The question, then, is which one will better serve your needs?

If you must choose between companies with equal distribution shares, consider what resources are necessary to get your invention to the marketplace. If your invention requires substantial financial resources to get it manufactured and established in the marketplace, then look to the company with those resources. Let's say that Company A and Company B have nearly equal distribution share, and both have the financial and manufacturing resources to produce your invention and get it to the marketplace. The next question is, do you want to be a small piece of a big pie or a big piece of a small pie?

The biggest licensee won't necessarily be the best.

Even though Company A may be the Goliath in your industry, your product may be diluted with hundreds of other products in its product line. Company B, with its smaller product line, may give your invention more personalized attention and feature it as a special new product.

WILL THE (SALES) FORCE BE WITH YOU?

The level of sales and marketing support your product will need to succeed is another factor to consider. When there are several steps in the distribution channel (i.e., manufacturer, wholesaler, jobber, retailer, and consumer), you are at the mercy of the talent of the sales staff at each step. It is typical for products to go through two or three tiers of distribution. This is especially true when they are distributed on a national or international scale. As the number of tiers of distribution increases, more responsibility is placed on the product to sell itself as knowledge of the product is diluted with each step. It is not uncommon for salespeople to become mere order takers, and this can spell death for a new product that needs real selling. Interviewing potential licensees as described later in this chapter will provide more information about this.

A MARKET-DRIVEN POSITION?

Some manufacturers produce goods for sale in several unrelated markets. For example, an injection molding plastics manufacturer that specializes in molding certain types of plastics and in specific sized molds may manufacture one item for the automotive industry, another item for the toy industry, and yet another for the hardware industry.

These manufacturers are sometimes referred to as *job shops,* or custom manufacturers. Although these job shop manufacturers may be in a position to manufacture your invention, they normally will only do so if you give them a specific order with specifications and conservative payment terms. These types of manufacturers are generally not good licensee candidates.

Other manufacturers concentrate on a specific market position. I call these manufacturers market driven. They are better licensee candidates. Rubbermaid is a good example of a market-driven manufacturer. Although Rubbermaid is an injection molding plastic manufacturer, it strictly manufactures products that it merchandises to specific markets, such as housewares. The company is also "vertically integrated."

It makes a product, distributes it, and markets it, from raw goods to retailer, right up the distribution channel.

Although both the job shop and Rubbermaid are manufacturers, and both are injection molders, Rubbermaid is a more appropriate licensing candidate for an invention geared to the housewares market.

One of the problems with searching databases such as *The Thomas Register* (a directory of U.S. manufacturers available in book form and on the Internet) for potential manufacturers is that it can be very difficult to sort out job shop manufacturers from market-driven ones. It's much easier to garner this kind of information from your interviews.

Making Contact

Once you've ranked your picks for potential licensees, it's time to start contacting them. The purpose of this initial contact is not necessarily to make a sale, but to do still more research, this time interviewing key decision makers at the manufacturers themselves.

The objectives of your initial conversation are to:

- Determine that you are communicating with the right company. This sounds obvious, but sometimes similar company names can cause confusion.
- Determine that you are talking to the right person—the one who makes new product decisions.
- Ascertain the company's level of interest in new products.
- Ascertain its level of interest in your specific type of invention.
- Learn its submission procedure.
- Ask questions and get as much information in as short a time as possible.
- Leave the company with a positive impression of you.

To achieve success, you must influence the people who influence the people who influence the decisions that you want influenced.

—ALAN GEIGER

FINDING THE DECISION MAKER

The key manufacturing people to contact are mainly found in the marketing department, such as the vice president of marketing, marketing manager, or product manager. In smaller companies the president or owner will be the one to talk to. In either case, you are looking for the person who makes the ultimate decision about which new products will be introduced into the company's product line. If this is handled by the patent department, you may want to have your patent attorney communicate with the company's patent counsel. Hopefully, though, you'll be able to avoid this expense, at least initially.

WORDS OF WISDOM

As a general rule, you should try to make contact with a decision maker in the marketing department, rather than the engineering department, when introducing your invention. Why? Because approaching the engineering department may be the kiss of death.

The marketing department generally thinks in terms of increasing sales for the company regardless of whether its own engineers or outside inventors develop the products. The mentality of the engineering staff is generally quite different, however. Remember that engineers are charged with developing and improving products and technologies for their company. Your invention may threaten their egos, since you thought of it first. This is unavoidable human nature.

Also, since engineers look for technical flaws when critically reviewing a new product idea or technology, their focus may be on what could go wrong with your product or why it would not work. Thus, you might receive a rather tough critique of your invention from the engineering department.

If you are going to end up in the engineering department, it is better if the marketing staff asks engineering to review your invention. The engineering critique will likely be less critical since someone in their own company is requesting the review. For all the engineers know, the new product idea may have originated from the president of the company. Therefore you may have a better chance of getting a favorable critique when someone within the company requests it than you would as an outside inventor with no contacts in the company.

What Decision Makers Can Tell You

Initial telephone conversations with decision makers continually amaze me. It is not uncommon to find the president of a major company taking an hour to talk to you about your product and the marketplace. Yet another key decision maker won't give you the time of day. It has been my experience that if you attempt to contact ten key decision makers, you will be able to reach five within a thirty-day period. Of those five, two or three will not be appropriate contacts, and you may or may not get referrals from them. One or two contacts will take the time to answer your questions about the marketplace and their company. Don't be surprised if you end up trying to get off the phone with them!

Listen and say no more than is necessary to keep the other person talking.

THE FIRST PHONE CALL

What follows is a great method for approaching companies with your invention. This requires excellent communication skills. Personalities, attitudes, and politics within a company will play an important role in your licensing efforts. If, after reading this, you feel uncomfortable with the process, you may want to have a licensing professional communicate on your behalf.

Call the potential licensee and ask to speak to the person who makes the final decision about introducing new products. When contacting a large company, this will probably be someone in the marketing department. Ask for marketing first and then ask who decides on new products. In small to medium size companies, you will generally speak to the president, owner, or vice president of marketing or sales. Very large companies may have a new product market committee, in which case you need to speak to someone on the product committee, preferably someone who likes to work with outside inventors.

Practice Makes Perfect—If it helps, write down and practice a script of your pitch before you place your call.

Once you have been transferred to someone, tell that person you are trying to reach the individual who makes the final decision about which new products are introduced in the company and ask if you have the right person. If not that person, ask who is. This is very hard for a novice. It is also extremely important. It is crucial that you speak with the key decision maker. If you can't get past this point in the interviewing process, get someone who can.

(Even if this is the furthest you can get personally, take comfort in the fact that you have already collected a lot of valuable market information. Doing so helps you make informed decisions when the time comes. You have also saved time for the licensing professional you end up enlisting.)

After reaching the right person, introduce yourself and explain that you are the owner of the rights to a new product or technology, and you are trying to find the right company to manufacture and/or market the product under a licensing (or other) arrangement. Without taking a breath, further explain that it is your intent to not divulge any confidential information at this time but to first find a company that is appropriate for further communication.

Then (still without taking a breath) describe the general category or premise of your invention without giving away any trade secret or confidential information.

When you've finished your pitch, listen carefully to the response. Following are several typical reactions and suggestions for how to handle them.

1. "I don't handle this type of product."

You may have reached someone too high on the totem pole or whose product area doesn't fit your invention. If so, ask for the name of the appropriate person, then ask to be transferred.

2. "Our company is not seeking any new products from outside sources at this time."

Ask if there are any companies that might be appropriate for your invention. (If they only volunteer one source, it can't hurt to prod for more.) Ask whether you can tell the other company who referred you. Sometimes it helps to have a company's name as an introduction. It would not be appropriate to name drop, however, if you do not want to be seen as shopping your invention to several companies.

3. "We only sell these products; we don't manufacture them."

This can mean different things. It could mean the company manufactures some of its products, but purchases others from another source. However, the actual manufacturer might be offshore and is not in a position to market your invention.

If they do not see the company as appropriate for manufacturing your product, ask whether the company would be interested in marketing your product if another company (its offshore manufacturer, for example) were to manufacture it. It may be that a joint deal of this nature is ultimately necessary to commercialize your invention. If, however, the company is simply not appropriate, then try to obtain as much information as possible about which companies are appropriate for your type of product or technology.

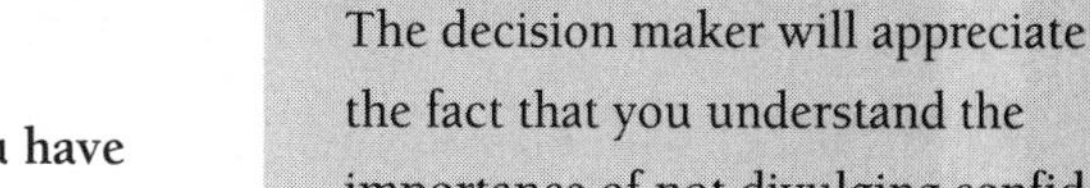

WORDS OF WISDOM

The decision maker will appreciate the fact that you understand the importance of not divulging confidential information because it could put the company in a vulnerable position.

4. "I will only consider your idea after you have signed the company's disclosure agreement."

Many companies will insist on this, and for good reason. Many corporations are leery of talking to outside inventors or their representatives because of the liability involved. One of a company's greatest fears is that an inventor will present a technology that its own engineers are already developing and the inventor will later claim to have developed it first. Inventors have won numerous court cases by falsely alleging that a company stole their invention.

If a company divulges its engineers' notes, revealing dates of development, to combat the allegations, the company could jeopardize its competitive position. Conversely, inventors have won court cases against companies that did not document their prior development in a suitable manner.

The result of all this is that companies have developed disclosure agreements for inventors to use so the companies can better protect

WORDS OF WISDOM

How do you know whether name dropping is inappropriate? When the key decision maker asks if you have presented your invention to other companies, this is a tip-off that that company or possibly that industry is fairly competitive and sensitive to this issue. It doesn't hurt to ask, point blank, whether this is a concern.

themselves. This is a common practice in most large companies and should not dissuade you from pursuing a company further.

5. "Sounds interesting. Tell me more."

This can be a very good sign; at least it shows the person's willingness to keep talking. In asking for more details, that person is generally trying to ascertain whether the technology would fit into the company's product line and whether it would be suitable for the company to manufacture. Don't divulge your date of inception or any confidential material at this time, and it is very unlikely that the person will ask you to divulge information that is confidential or proprietary.

EVERYONE HAS AN OPINION

When you are talking with decision makers at two or three different companies, ask their opinions of the other companies. Normally, they are more than willing to critique other companies in their industry. If two or three credible companies all speak badly of one particular company, you may decide against doing business with it. It is not uncommon for decision makers to pitch their company as they criticize the competition. It can be a refreshing change to have a company pitch itself to you instead of you having to pitch yourself to the company.

After you have broken the ice with your initial pitch and determined that the executive will spend a few minutes with you, by all means, take advantage of it. Here are the kinds of things you can learn from your conversation. If it helps, keep the book open to this list as you talk, with pen and paper handy to write down pertinent information. Don't try to barrel through the whole list; just use it as a guide. In fact, before you make your initial call, you might highlight the three or four questions that are most important in case the call has to be kept brief. The most critical questions are starred.

Ask these questions to garner market information:

- What drives sales in this product line: price, features (which ones?), packaging?*
- What are the ideal sales outlets: retail, mail-order, industrial?*
- What is the general retail price range for a product like yours?*
- Does the person have any idea of the potential sales volume for your invention? Would a ballpark figure here be five thousand, ten million, a hundred million?*
- What would the person see as the greatest resistance when introducing a new product like yours?*
- What company(s) offers the stiffest competition to this product line?*
- Would an invention like this support the company's market position?*
- How stable is the market, and what factors may cause it to fluctuate?
- To what degree does the company market its products internationally, and which countries is it strongest in? Where does it see the greatest competition coming from foreign markets?

- What would the time frame be for introducing your invention? How long would it take to introduce and establish your product in the marketplace?

WORDS OF WISDOM

Keep your contact talking and build a rapport with them. Laying the foundation for a good relationship in this first conversation will allow you to call back and gather more data in subsequent conversations. Remember, this initial call is to obtain as much information as possible about the company; it is not the time to make a heavy sales pitch.

Ask these questions about working with others:

- What has been their experience with outside inventors? How many outside inventors have they worked with in the past?*
- Has the company ever paid inventors for unpatented ideas, and approximately how much (ballpark figure)?*
- Do they sign an inventor's confidential disclosure agreement? Do they have their own agreements?* (See more about this later in the chapter.)
- What royalty rates are typical for the industry: 1 percent, 5 percent, 10 percent?
- Are up-front cash payments typical, and if so in what amounts?
- Does the company ever pay inventors consultant fees?
- Has the company ever maintained an exclusive license with minimum performance standards? Has the company ever had to relinquish a license because it did not maintain the minimum performance as outlined in the licensing agreement with the inventor?
- Has the company ever paid to patent outside inventions in the inventors' name?
- What are the names, addresses, and phone numbers of any outside inventors that they think you should talk to about the company?

Ask these questions about the company's manufacturing capabilities:

- What experience does the company have with engineering, manufacturing, and controlling the quality of the products that it manufactures?*
- What experience does the company have manufacturing products similar to yours? Where are such products manufactured?
- What might the company project as a manufacturing cost? What price might it charge its customer? What additional material does it need to ascertain the manufacturing cost?

Ask these questions about the company's background:

- What are the company's strong suits: marketing, manufacturing, engineering?*
- What market position has the company built for itself: superior value, low price, highest quality, superior service?*

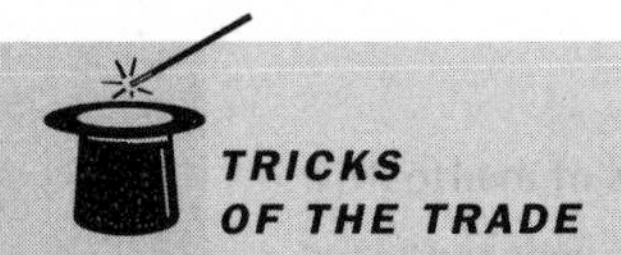

Let the Company Cost It for You—When you submit your invention to a manufacturing company for consideration, it will estimate its manufacturing cost based on potential sales volume. Having the company determine the manufacturing cost saves you time and gives an accurate result, much more accurate than you could obtain by costing out all the aspects of production at various machine shops or custom manufacturers.

- What has its success rate been in introducing new products?*
- How many new products has it introduced in the last year, last five years, last ten years?
- What importance does the company place on patents? Has it ever attempted to protect its own patent or its licensors' patents?
- How long has the company been in business, how long has the company operated under current management, and are any management or ownership changes anticipated?
- What departments or what people within the company need to be influenced to acquire the invention? Is there one area that may need more convincing?

Don't forget to ask what type of information people would like to see from you if they critique your invention. For example, do they want a brief description, complete documentation including engineering information, a sample of the invention, a working prototype, or a patent?

TOTAL VALUE

Your first conversation with a manufacturer will help clarify the value of your potential contribution to any given company. For example, you may have an idea for an invention without having done any engineering, test marketing, or product testing. Or you may have fully developed your invention, with engineering drawings and test marketing in local stores. Different companies in different industries place varying degrees of importance on these contributions.

Generally speaking, the more you have developed your invention, the greater value you will be to a company. If your idea is worth a certain royalty, your idea with a prototype may be worth more, a working model more yet, limited production with a test market still more, and a positive track record with actual sales experience may yield an even higher royalty.

When a company's strong suit is manufacturing rather than marketing, it may be less interested in the work you've done to engineer your invention and more interested in test market results. Another company's strength may be marketing, with a minimal engineering department. In this case, offering them a prototype may be particularly helpful. It is important to recognize how you can make up for a company's particular weaknesses. This is why it doesn't pay to spend a lot of your own money on patenting, prototyping, or test marketing until you have researched the market and determined what's needed to fill in the gaps. A large company that is fully capable of doing its own engineering development and test marketing may not be interested in what you have done to develop your invention. Such a company may prefer that you present a patented concept and leave the development to them.

Appreciate the value of leveraging.

Disclosure Agreements

If your potential licensees are large companies, or if any key aspect of your invention is not patented, then you will likely have to deal with disclosure agreements, either one of your own, or one of theirs. Disclosure agreements vary greatly in size, content, and format. Following are a few key elements that differentiate them.

WORDS OF WISDOM

Some companies are less interested in patented inventions and more interested in patentable ideas. The distinction here is that they prefer to use their own patent attorneys to afford the broadest patent protection possible for an invention. Information about their patenting position is something they will share with you when they think you have something that could contribute to the company's bottom line.

DOES THE COMPANY HONOR CONFIDENTIALITY?

Most company agreements will not offer to keep your invention confidential. This is not because the company is intending to broadcast information you give it, but rather is to protect the company against things it cannot control. For example, two different inventors may disclose nearly the same technology to the company. If one inventor broadcasts information about the technology and the other inventor asks for confidentiality, it could be construed that the company divulged this information to the second inventor when this was not the case.

Although many companies require their employees to sign confidentiality agreements when being hired, companies cannot completely control their employees and what they say. Such a confidentiality agreement could lead to a company being held unduly liable if information was leaked.

You have to rely on your own intuition and verbal commitments from the key decision maker you are dealing with. If that person assures you it is not company policy to divulge information disclosed to them by inventors and you trust them, then there is probably little harm in signing their disclosure agreement.

IS THE COMPANY WILLING TO ACCEPT CONFIDENTIAL INFORMATION?

Even when companies are not willing to guarantee they'll keep your information confidential, they may be willing to accept confidential information. There are two schools of thought about how you should handle this. The first is to divulge as much information to them as possible because the more information you divulge, the more indebted they may end up being to you. For example, precedents set by court cases suggest that in the absence of any other agreement, when you divulge information such as a description of an invention to a company, it is assumed that you did so expecting reimbursement if the company were ever to exploit the invention. It is assumed that you are not simply giving away your invention when you submit it to companies; it is assumed that you expect something in return. However, if this is your strategy, you should consult your patent attorney regarding the latest case law in this area.

The second school of thought is to divulge only what is necessary to get your point across. This reduces the chance of your confidential information being divulged to other sources. This decision is up to you and your sense of the competitive nature of your market, the company's reputation, and the nature of the information you are divulging. A happy medium is to be half wrong/half right: release need-to-know information initially, then follow up later with a complete disclosure if your strategy suggests this is a benefit.

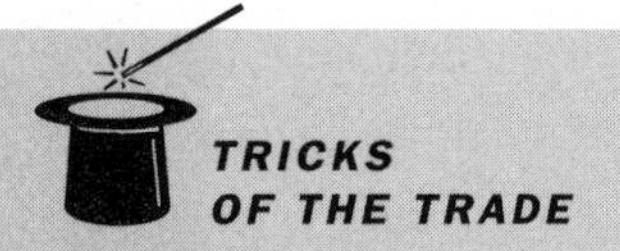

Bareing It All Can Keep You Covered—Sometimes it is beneficial to reveal every possible variation of a technology to a company after it has signed a confidential disclosure agreement. This can add to your stable of proprietary property because it anticipates possible improvements. Better that you take credit for an improvement than wait for the company's engineer to think of it and take credit instead.

MUST YOU RELY ON PATENT RIGHTS ONLY?

Most company disclosure agreements will suggest that when you disclose information to the company you are relying strictly on any right that you receive from a U.S. patent to receive remuneration from the company. Some major companies go a step further with their submission agreements, suggesting that you must rely on your defendable patent rights. They make it very clear that any patent submitted by an outside inventor must be able to withstand scrutiny and hold up under litigation.

In other words, just because you received a U.S. patent doesn't mean that you will still have it after a tenacious company is finished with you. You must have a patent that is iron clad. Some large companies reserve the right to reduce your patent to shreds if they believe there is a question about its validity. If all patents were put to this test of validity, many would fail. A corporation can pay a fee of a few thousand dollars, request a re-examination, or file a declaratory judgment suit to have your patent reversed. This is a nice way of saying "exterminated, no longer in existence, poof, up in smoke, no need to worry about maintenance fees."

With such a clause included, the key question to ask is whether the company has ever paid outside inventors for patented or unpatented ideas. Or has it ever caused an inventor's patent to be invalidated?

WHAT IF YOU DON'T WANT TO SIGN THEIR AGREEMENT?

In most cases company disclosure or submission agreements favor the companies more than they favor the independent inventors. If your invention is patented and you feel it is well protected under the patent, it may not make a big difference whether you sign the company's disclosure agreement. The trick is getting the company to review your invention without signing its agreement.

One way to do this is by explaining that your sole interest at this time is to determine whether the company would be an appropriate one for further communication. You want to know whether to continue with them or whether another company would be a better fit. Explain that your invention is patented and you only want to reveal information that is available in the public domain. This means you are only going to reveal a patent, which is public information and could easily be found in the Patent Office. In many cases, when you explain that you will send only the patent and no other information, the company will not request you to sign its disclosure agreement. In addition to keeping you from being unduly bound by an agreement, this hastens communication. Instead of waiting a week or two to receive the disclosure agreement in the mail, having your patent attorney review the agreement, then returning it to the company, you can fax or mail a copy of your patent with a brief cover letter to the key decision maker and be done with it. And if this company is ultimately not the right licensing candidate, you'll be able to approach another company sooner.

Your Own Disclosure Agreement

Before you submit your invention anywhere (really, before you show it to anyone in the industry), you should have your own confidential disclosure agreement. Such an agreement allows you to disclose your trade secrets, ideas, working knowledge, know-how, and other confidential material to a third party and to expect the third party not to reveal this information to others, or at least not to others outside of the company.

Although most major companies will not sign an inventor's confidential disclosure agreement, and in fact will insist you sign their disclosure agreement, there are many companies that are willing to sign yours. Since the key decision maker in most small to medium size companies is the company president or owner, it is easier to get that person to sign your disclosure agreement without extensive review by patent counsel. Also, these companies are less likely to have been the target of lawsuits and would therefore be less guarded.

In some ways a confidential disclosure agreement can offer you even greater protection than a U.S. patent. I have seen companies refuse to compete against an inventor after the inventor has licensed an invention elsewhere, even though the company could devise a way to work around the inventor's U.S. patents. The company refused because the inventor had revealed trade secrets and know-how under the terms of a confidentiality agreement that it felt compelled to honor.

I have seen this personally. I can remember one invention for which over two dozen companies had signed disclosure agreements. Half of them were major manufacturing companies, and the others were distributors. Those companies that signed the agreements collectively represented over 95 percent of the entire marketshare in that industry.

The disclosure agreements gave the inventor even broader rights than one would find with patent protection. The agreement declared that any designs, trade secrets, know-how, and improvements developed by the inventor and presented to a company would be the property of the inventor. The companies also agreed not to pursue development of the invention without the inventor's express written permission. Of course an attorney could argue that such an agreement could be challenged, but many companies would rather honor an agreement than hassle with a lawsuit. It ultimately depends on how much is at stake.

SHOULD YOU USE AN ATTORNEY?

It can be a disadvantage to have your attorney submit a confidential disclosure agreement to a company. This is because attorneys typically prefer to communicate with other attorneys. If an attorney submits an agreement to a key decision maker in a company, that key decision maker may be predisposed to having the company's attorney review it before the decision maker looks at it. This can cause delays, including time spent negotiating your disclosure agreement. Not to mention costing you attorney's fees.

I'm not suggesting that it is never appropriate to have your attorney involved in the signing of agreements on your behalf. In complex situations it would be important to have your attorney represent you every step of the way. This depends on your technology's overall potential and your overall strategy. It is a matter of cost versus risk, and I'm sure by now you're getting the idea that every situation is unique. Keep evaluating cost versus risk as you proceed through your invention project. It may help to consult with an invention licensing agent or broker about such issues. Paying a few hundred dollars of an hourly rate may cost you less than a mistake.

LEVELS OF PROTECTION

There are varying levels of protection available when submitting your invention to companies. Some of the possibilities are listed below starting with the generally preferred level of protection:

- A broad utility patent (and possibly a trademark) and a confidential disclosure agreement to cover any nonpatented aspects of your technology
- A utility patent only or a confidential disclosure agreement that recognizes your rights to your invention
- A nonconfidential disclosure agreement that recognizes your rights to your invention
- A letter to the company describing the invention you are submitting and suggesting that you want remuneration if the company commercially exploits the invention, and requesting a written response that the company accepts, or at least does not reject, the terms of your letter.

The above methods of protection all include a written response from the company. If you really want to have your invention evaluated by a company that is not willing to give you a formal written response, the next best thing is to document the information you send to that company and that the company has received it. You can request an acknowledgment letter from the company indicating that it has received information about your invention or request a receipt from the post office. If you are lucky enough to be there in person, have a receipt ready for someone to sign that describes your information and indicates that the person understood your invention.

THE BASICS OF A DISCLOSURE AGREEMENT

If you are going to have a close, ongoing research and development relationship with a company, a detailed agreement may be necessary. I've seen twenty-page confidential disclosure agreements. However, for the purpose of simply submitting inventions to companies, I have found simple agreements of a page or less to be much preferred. The sample confidential disclosure agreement included here is a culmination of my years of experience submitting inventions to companies.

I never cease to be amazed by the number of presidents and vice presidents who routinely sign this document. Its simplicity and informality make it less threatening, so key decision makers are more likely to sign it without having their attorneys review it and change it all around. The straightforward lay language

Disclosure Agreement

Beginning _____[date]_____, _____[Your name]_____("INVENTOR") and/or his/her agents or assigns will disclose know-how, patented and/or unpatented inventions, designs, trade secrets, etc. ("INFORMATION"), which INVENTOR considers to be a valuable commercial asset, relating to:

[Fill in your subject area; the more specific, the greater the chance for acceptance.]

The purpose of the disclosure is to allow confidential disclosure and communications between _____[Company's legal name or individual, if appropriate]_____ ("COMPANY") and INVENTOR to discuss the development, marketing, and other relevant issues with respect to said INFORMATION. This is not an offer to sell or license.

1) It is understood that no obligation, express or implied, is assumed by the COMPANY unless and until a formal written contract has been entered into, and the obligation of the COMPANY shall be only such as is expressed in the formal written contract. INFORMATION disclosed to COMPANY shall remain property of INVENTOR.

2) COMPANY agrees that the disclosed INFORMATION will be held in strict confidence, and INFORMATION will not be disclosed to any other persons outside COMPANY, or used by COMPANY, without prior written consent from INVENTOR. INFORMATION released within COMPANY shall be on a need-to-know basis.

3) COMPANY shall not be bound to secrecy when COMPANY can document that said INFORMATION was: 1) previously developed by COMPANY, its divisions, subsidiaries, or affiliates, or 2) disclosed to COMPANY by a third party completely independent of this or prior disclosures by INVENTOR, or 3) if said INFORMATION is found in the public domain.

These terms are agreed to by:

____

Signature Date

____

Name Title

____

Company

See Appendix F for a clean version of this agreement that you can photocopy and use.

makes it understandable for even the most skittish company executive. Also, the informal, fill-in-the-name lines seem to add to its nonthreatening flavor.

All the key ingredients are included. I've had several patent attorneys review this agreement, and they agree that it offers the inventor fairly broad protection.

This confidential disclosure agreement starts off by describing the purpose of the communication between you and the company—that is, to discuss your invention. This is stated in a friendly way. Then the agreement explicitly states, "This is not an offer to sell." If it were ever construed that you were offering to sell your technology or sell your invention to third parties, it could start the one-year time limitation in the Patent Office and adversely affect your opportunity to obtain foreign patents.

Some patent attorneys have assured me that when you offer to license or sell your technology to a company but not necessarily the products themselves, this is not construed as an "offer to sell." It would not affect your time bar until the company sells your invention in the marketplace. Other attorneys have told me that this is a gray area. Therefore, to be safe rather than sorry, I include this disclaimer regarding the sale of an invention whenever appropriate.

Next, clause 1 assures the company that it is under no obligation as a result of receiving information from you. This is important to companies. It means that you do not expect a company to be liable for reimbursing you unless there is a formal, written agreement between you and the company. Then a statement follows that any information submitted to the company will remain your property. This is the clincher. The company is agreeing ahead of time that the invention is yours and not theirs.

In clause 2 the company agrees to hold the information about your invention in strict confidence. It is important to follow this clause with exceptions, which appear in clause 3. Clause 3 specifies the terms in which the company shall not be bound to secrecy. This clause adds to the company's comfort level. As we have said, many companies are concerned that an outside inventor will submit an invention that has already been developed in the company. They are also concerned about the possibility that two inventors will come to them independently and submit the same invention. It happens.

Clause 3 helps to relieve these concerns. The exceptions in clause 3 are the reason for titling the agreement a *disclosure agreement* and not a *confidential disclosure agreement*. If the company falls within the exceptions listed in the third clause, the agreement is no longer confidential in nature. Nonetheless, even if a company is no longer bound to strict confidence as a result of the exceptions in clause 3, it has still agreed that the information about your invention shall remain your property. If two different inventors outside a company are making claims for the same invention, it is up to those inventors to duke it out with respect to which one of them was the first to invent.

Some inventors worry that a company will change dates on documents to suggest that it invented the invention prior to the inventor's disclosure to the company. The fallacy here is that the company doesn't know your original date of invention. Your documentation of originality could date back ten years! A company wouldn't know how far back to manipulate its dates. I have never had a problem with this. In fact, I've never experienced it at all. Remember, you might have a hundred-page disclosure document, and someone in the company could still tip off Cousin Benny in New York about the invention, and you would be hard pressed to prove the connection. The bottom line is, these agreements are no better than the integrity of the people in the company signing them. This is why it is of utmost importance to do your homework on a company before submitting your invention.

Most Fortune 100 companies won't even consider signing an independent inventor's confidential disclosure agreement. The larger companies will generally allow you to submit an invention only after it has been patented or if it has a patent pending, or after you have signed their disclosure agreement in which you agree to rely only on the rights of your U.S. patent or any other legal right you have.

Even if a company is not willing to sign your confidentiality agreement, it is essential that it at least recognize that whatever you present to it shall remain your property, that it will not use your invention without a written contract, and that you expect payment if the company does use it. You should at least state this in the cover letter that accompanies your invention submission.

Patent Strategy, Continued

Why do most major companies require that an invention be at least in a patent pending stage prior to submission? Your invention must be fully disclosed in writing when you apply for a patent, so a pending patent assures the company that there is specific documentation of your invention. Your patent application can clear up any question or dispute over exactly what you invented as compared to other developments already in existence in the company.

Up to this point, it may have been strategically beneficial not to have filed a patent application. Patent applications are costly, and the process takes time. Plus it's helpful to interview companies first to confirm that your product is commercially viable. However, if the companies that seem to be appropriate licensees require a patent before they evaluate your invention, this may be the time to start a patent application.

Your Next Steps

Only share the serial number and date of your patent application with your patent attorney or patent agent. In addition, only share the claims of your patent with your attorney, agent, or licensing professional, who needs to know the claim details for marketing purposes. This information could be used against you if a hostile company wants to circumvent your patent.

By now you have made contact with key people on every distribution level in your field, including manufacturers and potential licensees. With this comprehensive market research in hand (and probably a higher-than-normal phone bill), you should have enough knowledge to either make informed decisions or be very confused. (If you're confused, just keep gathering information until you find some clarity.)

Do you still think you have a licensable invention? Are you comfortable with your prospective list of licensees? Have you picked a first candidate to offer your invention to?

If manufacturing your own invention is the way you're leaning, you can either pursue manufacturing on your own or first submit your invention to manufacturers for the purpose of picking their brains about the viability of your product. The latter is a very aggressive ploy, but a manufacturer's knowledgeable critique of your invention could help you enhance your product. And, if the manufacturer signs your disclosure agreement (in which you've written language that restricts the manufacturer from competing against you with a similar product), then you've not only succeeded in obtaining valuable information about your product and the market, you've also put yourself in a better situation by restricting (or possibly eliminating, depending on the language in your agreement) the competition.

The down side of this aggressive ploy is that it can backfire. The manufacturer may decide to introduce a product similar to yours because you showed it how much market potential there is. You show your hand, so to speak, and the manufacturer designs around you. If you can submit a basic concept without revealing your trade secrets, then it's probably okay to do some brain-picking. Obviously this is a situation that must be carefully weighed.

If you decide to start your own business instead of pursuing licensing, the marketing section of your business plan is mostly in place with all the market research you have done. If you later decide to sell your company or license your invention after it has earned a positive sales record, the contacts you have made will put you in a good position to hit the ground running, no matter what your choice. In the end all of your hard work will be worthwhile.

You may have learned that you need to rethink your commercialization strategy altogether or start researching a slightly different industry segment. If licensing is the route you want to go, then you've arrived at yet another decision point. Is what you have what manufacturers want? You may have spent all of your time securing a patent when the manufacturers are only concerned about the potential market. Or vice versa—you may have test-market data to validate that your product will sell, but be weak on the patent side. If these types of gaps exist, someone will need to fill them, either you or one of your licensing candidates.

You may now choose to further develop your invention to enhance the features preferred by the industry, or you may now decide to apply for your patent. If you have the green light to submit to one or more companies, either for the purpose of further market research or to license your invention, the next chapter will guide you through the process of developing your commercialization strategy.

THINGS TO REMEMBER AND CONSIDER

- Select manufacturers carefully.
- Follow up diligently without being overbearing.
- Listen and follow the advice of sages.
- Maintain flexibility when it comes to confidentiality agreements; don't get bogged down here.
- Find out what people in your industry want in a new product.
- Continue to reevaluate your position and that of the people with whom you are working.

New Submission Resource

DIMWIT's Guide for Inventors (www.DIMWIT.com) is a new online resource that helps inventors prepare relevant information about their invention to submit to companies. Simply follow the prompts on the website and answer the questions in the tutorial guide. DIMWIT will automatically create a professional presentation of your invention.

DIMWIT allows you to describe the benefits, advantages, and general features of your invention without disclosing your trade secrets or any confidential information. This approach enables you to break the ice with companies when neither of you agrees to the other's disclosure agreement. Plus, DIMWIT focuses on the pertinent information that companies want to know, thereby increasing your chances of getting to the next step.

CASE STUDY

The Picture Perfect Deal? (Part 1)

One of my clients was a prominent California inventor, holder of several patents. Two of his patents relate to technology supporting a new generation of laboratory and scientific heating devices, an electric Bunsen burner of sorts.

This invention utilizes a heating element, similar to a light bulb, and an ellipsoidal (half sphere) reflector that reflects the heat and light emitted from the heating element to a point in space. In this case, it concentrates at a point about a foot above the reflector.

This offers several advantages. Since there is no open flame, it can be used in organic labs and other applications where open flame is not permitted. It also allows you to heat in specific areas within a test tube or beaker without having to turn the test tube on its side. It won't replace existing Bunsen burners; however, it is a new alternative for applications that were previously impossible.

See the following cover page of the inventor's U.S. patent.

My company accepted this project into its Market Evaluation and Commercialization Program. Using the methods outlined in this book, I backtracked through the channels of distribution, starting with end users for such a device.

I first went to a university laboratory where burners are used and met with a purchasing agent. I learned that there was nothing else like this product on the market, and the only other available alternatives were electric hot plates and similar devices that use conduction heaters to warm beakers. None of these could perform the function of tube bending like the inventor's apparatus.

Through field market research, I learned that there were three major distributors of scientific equipment in the United States and that these distributors controlled well over 90 percent of the marketplace. I got the names of the distributors' sales representatives from the purchasing agent. Telephone interviews with the reps confirmed that two companies controlled most of the distribution in the marketplace, with a third company representing a lesser share.

Although the top two distributors had very different marketshares, they had similar shares of distribution. In other words, they both sold product to most of the same places. The leading distributor in the marketplace was also the most active in overseas sales. The distributor has 3,000 employees and 1,200 suppliers. It offers a line of 130,000 different products and manufactures 25 percent of the items it sells. If these people weren't familiar with the market, no one was. The rep gave me the name of a company decision maker, the vice president and general manager of the laboratory equipment division.

The vice president was very agreeable when I called him. He was willing to look at and critique my client's invention. He also gave me an overview of the company and indicated

United States Patent [19]

Downs

[11] **Patent Number: 4,739,152**

[45] **Date of Patent: Apr. 19, 1988**

[54] **ELECTRIC RADIANT HEATING DEVICE FOR LOCALIZED HEATING OF OBJECTS AND SUBSTANCES**

[76] Inventor: **James W. Downs**, 3846 Dunford Way, Santa Clara, Calif. 95051

[21] Appl. No.: **782,031**

[22] Filed: **Sep. 30, 1985**

[51] **Int. Cl.⁴** **H05B 1/00**

[52] **U.S. Cl.** **219/347**; 219/349; 219/354; 219/358; 219/461; 313/113; 313/114; 362/298

[58] **Field of Search** 219/339, 342, 343, 346, 219/347-349, 358, 461; 313/113, 114; 362/298

[56] **References Cited**

U.S. PATENT DOCUMENTS

2,059,033	10/1936	Rivier	313/114 X
2,512,061	6/1950	Huck	219/347 X
2,785,623	3/1957	Graham	219/349 X
3,253,504	5/1966	Vollmer	313/114
3,401,256	9/1968	Siegla	219/354 X
3,427,433	2/1969	Foreman et al.	219/354 X
3,427,435	2/1969	Webb	219/349 X
3,434,818	3/1969	Chauvin	219/349 X
3,621,198	11/1971	Herbrich	219/349

FOREIGN PATENT DOCUMENTS

502262	7/1930	Fed. Rep. of Germany	219/349
2506494	8/1976	Fed. Rep. of Germany	219/347
762484	1/1934	France	362/298
1089131	9/1954	France	219/347
359778	6/1938	Italy	362/298
532914	2/1941	United Kingdom	219/347
727346	3/1955	United Kingdom	219/349

OTHER PUBLICATIONS

"Radiant Heating Using an Ellipsoidal Reflector"; B. Uavala, A. Muries; Journal of Physics E, vol. 7, No. 5, pp. 349–350; May 1974.

Primary Examiner—Anthony Bartis
Attorney, Agent, or Firm—Warren H. Kintzinger

[57] **ABSTRACT**

An elongated, coiled, electrical resistance heating filament is positioned along the major axis of an ellipsoidal reflector with its inner end positioned at the inner focus of the reflector and with the coil extending outwardly toward but short of an annular rim defining the front of the reflector. The coiled filament is supported by a pair of power supply leads connected to the inner and outer ends of thereof and forming stiff conductive support pillars mounted in a thickened support and seal portion at the apex of the reflector. The front of the reflector may be closed by a glass plate sealed to the annular rim. A secondary reflector may be provided on the major axis of the ellipsoidal reflector between the outer end of the coiled filament and the annular rim. The secondary reflector may be a figure of rotation of any shape, e.g., parabolic or conical, capable of throwing energy that would otherwise escape back through the coiled filament and serves to conserve energy and shield the eyes of the user.

11 Claims, 2 Drawing Sheets

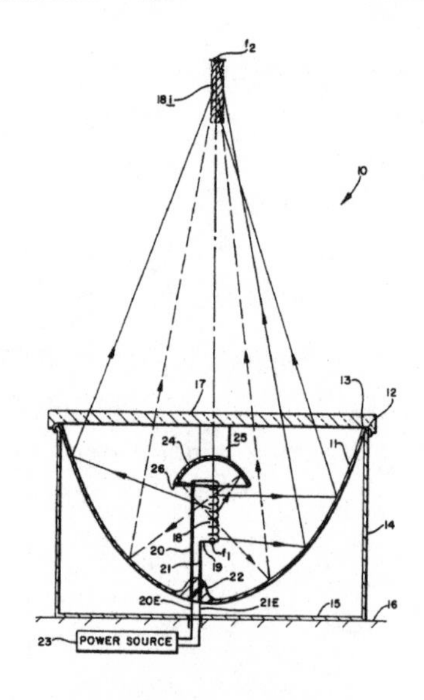

that the company was rapidly expanding in Europe to add to its stronghold in North America. I sent him a brief one-page letter, a copy of the patent, and an informal illustration provided by the inventor.

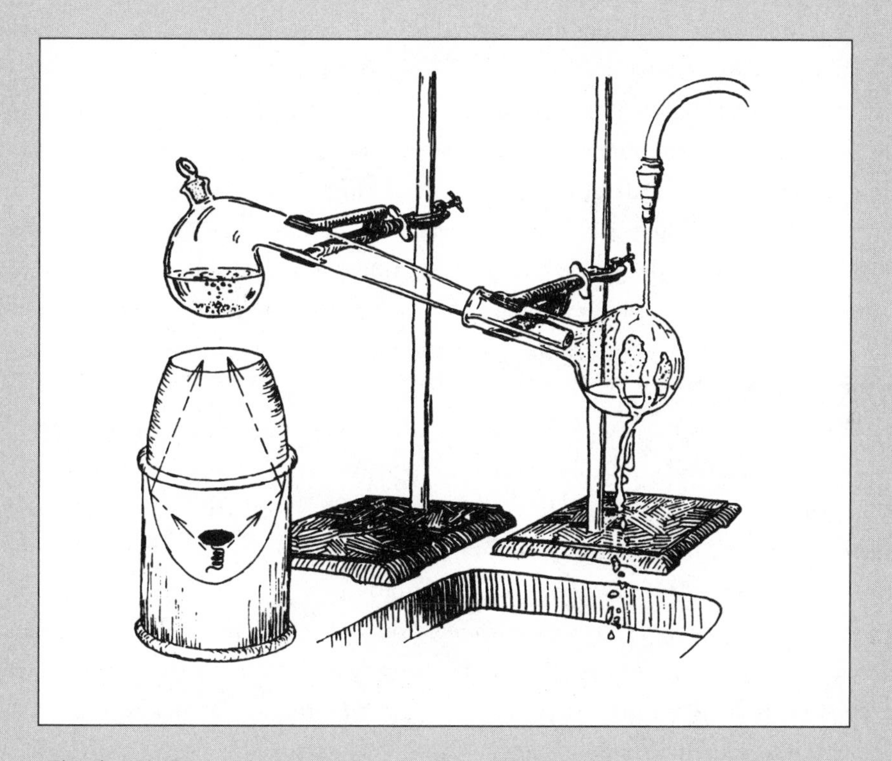

The electric Bunsen burner projects an inverted image of a tungsten filament at the second focus of an ellipsoidal reflector. The image of the filament serves as the flame area of a conventional gas burner and attains a temperature nearly as great as that of the filament. The advantages of an electric Bunsen burner over conventional gas burners are: no open flame, electricity is more available than gas, no combustion products, the burner can operate inside a bell jar, and the precise and measurable energy can be controlled with an SCR or Variac.

While waiting for the vice president to review the invention, I ordered catalogs from four different distributorships, two major ones and two smaller ones. I also ordered brochures from several of the companies that manufacture Bunsen burners and electric heating devices that I had identified during my initial market research, and from companies the manufacturer's representatives referred me to. About six different companies were manufacturing and marketing a product line in this general area. Two of these companies were based in Europe, one in England and one in Germany, although they had headquarters in the United States as well.

It was a lot harder reaching the vice president the second time around. I finally reached him two months after the submission, which he'd just had a chance to review. He said the invention was "more interesting than originally thought," but he was concerned about safety because he could not determine whether the heating area was visible or not. His next step was to have engineering and marketing review the invention. He asked that I call him six weeks later.

I finally reached him again three weeks after that. He was quite enthusiastic about having the invention in the company's product line. However, because of a backlog in its manufacturing facilities, and because the item was not considered "hot" enough (from a sales perspective), it was not in a position to manufacture the item. The company was very interested in distributing it, and the vice president was more than willing to refer us to the manufacturer he thought would be the most appropriate. The manufacturer he recommended, based in Chicago, was the world's leading manufacturer of Bunsen burners, in business since 1908.

My initial contact with the president/owner of the manufacturing company in Chicago was not entirely promising. He was just getting ready to leave on a two-week vacation and was not anxious to have a lengthy conversation with me. I briefly described my client's invention, and though he said the company had "bigger fish to fry," he invited me to call him back after he returned from vacation.

In our second conversation, he was much more relaxed. He felt his company was well positioned to manufacture my client's invention, and he asked to see a copy of the patent. He wanted to determine the manufacturing cost to ensure that the product could be affordably marketed. Within a month, he had shared the patent with his chief engineer, who said that it looked "reasonably interesting." It was, not surprisingly, another month before I reached the engineer for a firsthand account of his evaluation. The engineer said that he felt the invention was a "viable alternative." I suggested that the inventor and engineer meet so the inventor could answer specific, detailed questions about the technical aspects of his invention. The engineer was in the process of completing a big deal with another inventor and wouldn't be free for another two months.

Don't assume that people receive what you send. I had to send this information twice; evidently it was not received the first time. This type of thing happens. Follow up within a week or so to make sure that your material was received. This is also a good time to get spontaneous or first-impression feedback about your invention.

WORDS OF WISDOM

The company in Chicago was a good lead, and the distributors' reference was a great boon. But rather than targeting just one manufacturing company, I decided to do a bit more market research by making telephone contacts with a few of the other potential manufacturers. I learned that these operations were fairly small and not necessarily inventor friendly. I also discovered that electric Bunsen burner products were taking some of the marketshare away from gas Bunsen burners. The additional research gave me a clearer picture of the marketplace and suggested I was already pursuing the best potential licensee.

Since this company was interested enough to assign an engineer to the project, and since there were few other companies waiting in the wings, the inventor and I concluded that it was worth sticking it out with this company to see how far we could get.

When the inventor finally met with the engineer (on the company's dime), they decided to work jointly to produce a prototype of the invention. Both would source materials to help estimate a manufacturing cost figure. Over the months they worked together, the inventor and engineer sometimes had opposing ideas of how the invention should be shaped. An unreasonable inventor might stick to his guns and not modify his design. Luckily my client was willing to conform somewhat to the desires of the company, and in doing so he found an even better way to configure his invention, all of which fell within the claims of his patent.

One day I received an exciting phone call from the inventor: "It works!" The prototype boiled water and bent glass tubing. He shared the results of the test in a letter to the engineer. Now it was time for me to rejoin the project and call the engineer to discuss the possible terms of a license agreement. We agreed that the final test of the technology would be performed at the manufacturer's headquarters in Chicago in about three weeks, after which the inventor, the company president, the engineer, and I would sit down to discuss possible commercialization. It was now more than twelve months since I had first submitted the invention to the company. My ultimate objective would be to leave this meeting with all parties in agreement on the basic terms of a potential license contract.

(To be continued . . .)

HINDSIGHT LESSONS

Patience can be a virtue. Even when everything seems to be going well, the licensing process can take much longer than you might expect. It never hurts to go the extra mile and acquire whatever market information you can. Sometimes the success or failure of a conversation depends on one person's frame of mind or workload. Sometimes busy people need a little prodding to keep the project moving; however, too much of this will pigeonhole you as an annoyance and hinder your project. Moderation and awareness are the keys.

You also need to be willing to rethink the configuration of your invention. Be attentive to the feedback you receive from the manufacturers; their ideas may enhance your technology or product.

Leverage worked here. The invention's value hadn't yet been proven with a working model, but the concept was simple and made sense. The major distributor's interest provided enough leverage to entice the manufacturer to pay for the prototype. This is an example of the importance of interviewing—the information we gathered allowed us to add value without an additional investment by the inventor. Our next step will be to leverage the results from the laboratory test to influence a licensing deal.

Nothing is invented and perfected at the same time.

—JOHN RAY

CHAPTER FIVE

The Plan Comes Together

Submitting to Companies and Forming a Commercialization Strategy

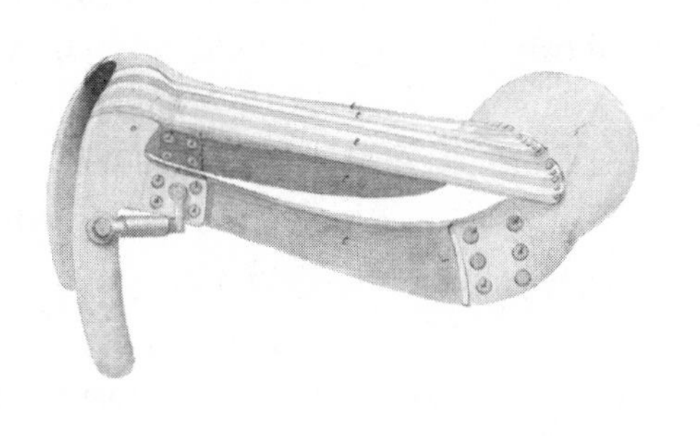

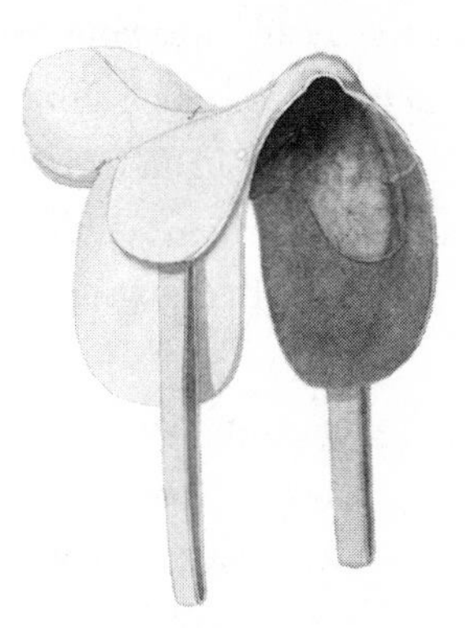

ONE MUST BE A GOD TO BE ABLE TO TELL SUCCESSES FROM FAILURES WITHOUT MAKING A MISTAKE.

—Anton Chekhov

You have now had a chance to thoroughly investigate the companies that you are submitting your invention to and have talked to the key decision makers.

Bear in mind that you are not yet out of the woods in establishing a profitable licensing deal. Although it may appear like licensing is the way to go, it's not ultimately up to you. What if no one ends up making you a decent offer? An entrepreneurial effort may be your only chance to introduce your invention to the buying public.

You will certainly have a better idea of this after following the steps outlined in this chapter. Although market research was the primary motive in your early conversations with the target manufacturers, now it's the secondary element. Getting a license or other suitable deal is your primary goal. Here's how to go about it.

A Few Words about Strategy

You'll see in this chapter, as with the "Smooth Success" story in Chapter 4, that there can be advantages and disadvantages to submitting to several companies before beginning negotiations with one. Your licensing strategy will depend on

Spring riding saddles by Richard Harrison (U.S. Patent No. 8011X), New York, New York, 1834.

several factors, including your level of patent protection and how closely competitive your industry is.

From the six companies you were initially considering, you will have narrowed it down to about three companies that have shown some interest in seeing your invention and together share at least 75 percent of the market in your field. Rank them in the order that makes the most sense to you, taking into consideration everything you've learned about each company's marketshare, reputation, values, and so on.

Let's say all three companies you are considering are equally prominent in the marketplace (or one is dominant and two are strong seconds), and all have shown interest in an invention like yours. There may be an advantage, especially from a time standpoint, to presenting your invention to all the companies at once. If, however, you believe the response from one company will enhance your presentation to the remaining companies, you may want to present to that company first.

An expert is a man who has made all the mistakes which can be made in a very narrow field.

—NIELS BOHR

There have been times when I submitted an invention to a company that turned out to be inappropriate for licensing, but benefitted from the company's feedback and enhanced the presentation I eventually made to a more appropriate company. This might happen if your invention has two or more useful purposes or applications. In contacting the first company or two, you may realize that one particular application is of little or no interest to the industry, yet the second or third application is of significant interest. Such knowledge will help you realign your presentation and steer it toward the right company, department, or subsidiary.

The amount of detail you decide to disclose about your invention will also play a role in your strategy of which company to submit to and when. If your invention is already patented and advertised, for example, on the Internet, then there may be little harm in contacting several companies at the same time to get their responses as quickly as possible. Likewise, if you can talk about your invention's basic advantages and features without disclosing how it works, even if your invention is unpatented and a secret, you may want to submit to several companies at once. But if your industry is very competitive, discretion may be your guide.

Deciding which and how many companies to submit to is critical. You are almost always better siding with caution and submitting to one company at a time. You'll have to gauge your specific situation and formulate your strategy accordingly.

What to Submit

When you had your conversations with company decision makers, they probably told you what type of information they wanted to receive from you. Most will be looking for similar information to initially evaluate whether your invention has a place in their company. Do a DIMWIT.com presentation, or provide:

- A brief description of your invention.
- A description of any specific manufacturing requirements.
- What about your invention sets it apart from existing technology, and why it is better.
- Your idea of who will use your invention (its market).
- What you expect from the company: for example, you want it to manufacture and market your invention and pay you a royalty or other fair remuneration. (Specific dollar amounts or exact terms of how much money would be paid up front are not necessary now. This negotiation will follow if the company is interested.)
- A list of the application(s) or uses for your invention.
- A copy of any patents.
- A sketch, photo, or sample of your invention if appropriate.
- Any documentation that verifies the workability or desirability of your invention, when appropriate.
- Any other backup documentation.

The information you provide the company at this point should be brief and concise. The basic information from the first six points should be only one to two pages in total, followed by backup documentation and sketches.

Don't get caught up trying to convince the company that your invention works and explaining how it works. Instead, show the company why your invention is better than existing products. If possible, describe what your invention does and what advantages it possesses without revealing its exact working nature or any trade secrets about how it is made.

The following is an actual submission letter I prepared for a client. It is a good example of how you can describe many features and advantages without revealing how your invention works.

Notice that the sample letter does not mention who could make use of the invention. This was done on purpose

How to Hook a Nibbler—A company that resisted signing any type of confidentiality or other agreement with you may change its mind when you present it with a description of the advantages and benefits of your invention. It may become so keenly interested in acquiring the rights to your invention that it will forego standard operating procedure and sign a special agreement honoring the confidential nature of your trade secrets. That is why a step-by-step approach can be quite useful when dealing with trade secrets.

April 10, 2009

Mr. Smith, President
Big Mouth Communication, Inc.
50 Spring Street
Latrene, NY 04746

Re: Mike Bayless's Proprietary Ringing Module Circuitry

Dear Mr. Smith:

It was a pleasure speaking with you recently regarding this new product opportunity. Please find enclosed the information you requested regarding Mr. Bayless's patented ringing module. Docie Marketing is Mr. Bayless's licensing agent. It is our goal to help facilitate the commercialization of his technology.

The electronic ring monitoring device for a telecommunications line is an interface that will allow the control of a 120 V high-current AC source by the ring voltage of a low-voltage telecommunications line. The circuitry completely eliminates the use of more expensive mechanical relays, which have been used for over seventy-five years. It provides for a longer lifespan, is less expensive to produce, and can be used in a wide variety of applications. The ring module is programmable, designed to work on the diverse types of telecommunications lines used today, and screens false rings. More detailed information is available in the enclosed patent and data sheet.

We are interested in knowing whether your company is in a position to manufacture and/or market this technology. We will contact you within the next two weeks to discuss the exciting possibilities that this revolutionary invention may provide for your company.

Thank you, Mr. Smith.

Sincerely,

Ronald Louis Docie, Sr.
President, Docie Marketing

cc: Mr. Mike Bayless, Pte.

Enclosures: U.S. Patent No. 5,509,068 and data sheet

because we were not sure of the answer, and we wanted to find out from the companies to which we were submitting the letter. Remember, you don't have to know it all. Let the companies share their expertise with you.

Following up on your submission is essential.

Send only copies of the supplemental material, not your originals, and document everything that you send out. Companies will sometimes return material to you upon request. Sometimes they copy your material for their files before returning it in case there is any dispute as to exactly what was submitted to them. If this is a concern, you may request in your initial submission that all of your material be returned to you upon rejection, and that the company not make copies.

When soliciting companies about licensing your unpatented invention, remember to indicate both in your conversations with them and in your follow-up letters that you are not offering your invention for sale; you are doing your initial technical and market feasibility research to determine your invention's value. The reason for this distinction is to avoid the one-year time bars we referred to earlier. Clearly stating in writing that you are not making an offer for sale at this time will help to prevent that.

Corporate Fortune Tellers—There are three key elements that affect the commercial success of your invention: manufacturing cost, sales price, and projected volume. Trying to estimate these figures is a chicken and egg situation; you often can't figure out one without knowing the other. Although it appears that this could possibly stalemate your communications with a potential licensee, you can make the confusion work for you if you're careful.

If you are in discussion with a marketing person who has little engineering or production experience, they may be able to calculate the potential sales price, but not how much it will cost to produce your product. They may ask you to provide this information to be able to evaluate whether or not to license with you. Unfortunately this is the person you are trying to get information from, and they're asking you for figures!

Since you need a rough sales projection to determine the manufacturing cost, this is an opportunity to learn something. Tell your marketing contact that you first need to know the potential sales volume. Such a projection can be based on whatever price the market will probably bear. Then you can ask your sources to establish a ballpark manufacturing cost and see if the cost will support the sales price with room for profit. Another option is to suggest that the potential licensee have its production department source the product. This saves you time, and it will probably get lower and more reliable cost figures in-house anyway. Also, companies like the idea of using sources they are familiar with.

Initial Responses from Key Decision Makers

After you submit your material, wait about a week and call your contact to make sure they received the information and to see whether there are any questions about it. Solicit any initial impressions, whether good or bad. If your contact needs more time to evaluate your invention, ask how long and offer to call back at that time.

WHEN THE RESPONSE IS NEGATIVE

In my experience, at least 85 percent of submissions receive an initial negative response—even when you've done your research well. It is not uncommon for the reason for rejection to make absolutely no sense. Follow up on this. It is of utmost importance to find out what about your invention was not seen as feasible, if for no other reason than to make sure it was understood.

First, ask for an interpretation of your technology and opinions about its benefits and deficiencies. Many times even key decision makers do not thoroughly understand or appreciate the exact nature of an invention. The rejection could be the result of a report from the engineering or marketing staff. I find that at least half the time a rejection was based on someone not thoroughly understanding the exact nature or intent of an invention. It needed to be explained in a different way.

The majority of invention submissions are rejected for several main reasons:

- It does not fit the company's market or positioning strategy.
- The company is not in a position to pursue an invention of this type at this time. In other words, its new product developers are tasked to the max.
- The company feels the projected price is greater than what the consumer will pay.
- The invention may be illegal or violates EPA, OSHA, or other laws unknown to the inventor.
- The company has already seen a similar invention in the trade.
- The company may have attempted to market a similar concept and met with failure for any number of reasons that would also apply to your invention.

THE WAITING GAME

Don't be too concerned when executives take four to eight weeks to respond to you. Remember, they have their own set of priorities, including responding to their superiors' requests, supervising the daily administration of their company, managing staff, developing existing products, strategic planning, and so forth. Evaluating a new product is usually last on their list of priorities. It's the nature of this trade. Calling the company to give them a periodic reminder is fine; constantly nagging spells death for your project.

In most cases an invention is rejected not because it is bad, but because it does not fit with the company's marketing and positioning strategy. Ask the key decision maker which companies would be more appropriate for you to contact. You may find that you need to contact an entirely different set of companies than you originally researched. You may have targeted a certain market niche only to find out that a completely different niche is appropriate for licensing. I've experienced this numerous times.

Talk directly to your contact to find out the details of why your invention was rejected and get a referral to other companies. Rarely will a company supply this type of detailed information in a letter. A standard rejection letter will say, "Thanks, but no thanks. It doesn't fit our corporate strategy at this time." Although a company won't divulge information regarding new internal developments in a letter, you can sometimes get some of this information from a telephone conversation.

If you never received any written verification of your submission to the company, you may want to ask for a written response, even if it is a rejection letter. This will at least give you written acknowledgment that the company received your submission.

When a company goes to the trouble of doing an extensive investigation of the merits of your invention, it is bound to make comments that would change the way you approach the manufacturing and marketing of your invention. Even if it has no further interest in the invention, you could still use this information to enhance your disclosure to the next potential licensee.

WHEN THE RESPONSE IS POSITIVE

Two of the more prevalent responses are: the company wants to do more investigation of the invention, or it wants to know on what terms you are making your invention available, or both.

THE COMPANY WANTS TO INVESTIGATE FURTHER

When the company wants to investigate your invention further, this is a chance to learn more about your invention's value. Ask what the decision maker likes about your invention; which features and benefits are most appealing? The answers will also verify that the key decision maker thoroughly understands the key features and benefits of your invention. If not, you need to provide more accurate information, first verbally and then followed in writing if necessary.

When it is clear that the decision maker understands your invention, find out the following:

- What is the internal procedure for ongoing review? Will your idea be sent to the marketing department, followed by engineering, or what?
- Who in the various departments will respond to the key decision maker?
- What is the anticipated timetable for completing the review of your invention?

Indicate your level of interest in communicating directly with the departmental heads and employees who will be reviewing your invention. Offer to answer their technical questions directly. For example, an engineering review may conclude that the current mode of manufacturing would not be cost effective for your invention. Based on your expertise, you may be able to suggest an alternative manufacturing method. Or perhaps the marketing department is concerned about the appearance of your invention, which would affect its acceptance in the marketplace. You could respond to this issue as well. You may work with these same people later to develop your product. This communication may serve as the beginning of an ongoing relationship between licensee and licensor.

THE COMPANY ASKS ABOUT YOUR TERMS

What terms do you want? In addition to determining your reasonableness, this question can be a very good sign. It indicates that the company has some degree of interest in pursuing your invention further. But it also puts the ball in your court, where you don't want it to be just yet. My advice is to lob it back.

Regardless of what you declare the worth of your invention to be, the company generally will already have an idea as to what it believes your invention is worth.

To avoid asking for too little, say "I want what is fair; what do you think that is?" This puts the burden back on the company to make an offer. Ask what the industry standard is, and confirm that figure later with a key person in another company in the same industry. Then double that amount, and start negotiating from there. It is generally a matter of negotiation after that, and may the better negotiator win.

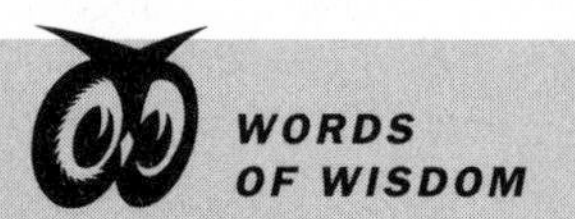

Looking for a cool million up front? Some inventors have requested a minimum $1 million cash payment to secure rights to their inventions. This is almost always unreasonable and far beyond what a market will bear. Companies will look for signs of reasonableness from inventors or their agents. If a company believes your demands are unreasonable, it may forego any further communication.

Determining Your Invention's Market Value

Knowing the total value for your invention is crucial to the licensing process. Ultimately the value of your technology and proprietary rights is what the potential licensee will pay. In some industries, fierce competition drives down the profits so little money is available for royalties. On the other hand, you could be on the cutting edge of a very profitable industry that has much to offer in this regard.

Here's how it works. The marketing department of the company you are dealing with will determine at what price the invention should be sold. The manufacturing department will determine the manufacturing cost. The difference between the two is the gross profit margin. The company then will consider the probable sales volume and product life cycle (longevity) of your product multiplied by the gross profit margin to determine the value of your invention.

It's easy to overestimate your invention's market value.

If you're dealing with one company at a time—not shopping your invention around—you must have a certain amount of faith in the accuracy and honesty of the potential sales figures the company develops. However, you can perform your own market research by hiring an independent market research firm or contacting trusted manufacturer's reps, distributors, and retailers for their insights.

Specifically you should try to ascertain the number of potential sales outlets for your invention and the average number of turns per year for this type of item. A *turn* is the number of times a store has to reorder your product. Match this information with the average number of pieces that the store may stock, and you can extrapolate the data. For example, if stores order an average of 12 units at a time and a reasonable turn in the store is 5 turns per year, that's 60 sales per year per store. If there are 10,000 possible sales outlets for this item, then your projected sales would be 600,000 units per year. Although such a figure is not exactly accurate, it establishes a benchmark.

Typically companies' projections are lower than inventors' expectations. If there is a large difference between your perception of market and the company's perception of market, then it may be beneficial to explain to the key decision maker that your expectations are higher and ask why his or her expectations are so much lower than yours. This is a reasonable question and shows humility.

If the company believes it can sell only 300,000 per year, you are prepared to ask why. Is the company only capable of reaching half of the potential sales outlets? Does it believe your invention will be a slower mover, i.e., with fewer turns, than what you anticipate? Does it think stores will stock a smaller number of units than you anticipated?

A dominant company in the marketplace, one that commands 80 to 90 percent of the marketshare for the entire market, will be in a good position to determine the value of your invention because of its experience. Besides its insight, it may also have some control over the market. After all, it already controls the marketshare.

Discussing these issues not only helps you understand the nitty-gritty points of distribution, it helps you see a licensee's point of view, which can expose the strengths and weaknesses of your licensee candidates. More importantly, it encourages potential licensees to visualize your item in the store along with their other products and to consider the potential profits. When they start to visualize your item, it begins to feel more like a real scenario and makes it easier for them to take ownership of the notion of licensing this invention from you. This psychology can help keep your project moving along in the minds of the key decision makers as they seriously consider licensing your invention.

PERFORMANCE STANDARDS

If offered the opportunity, many companies would prefer to have exclusive rights to an invention with no guaranteed minimum sales performance. An obvious disadvantage of this is that once the company has the exclusive rights to commercialize your invention, it could produce only one or two and pay royalties on that amount and sit on your invention thereafter. Without certain sales performance standards in the license contract, you would have no recourse. So when a company indicates that it is interested in your invention and wants the exclusive rights to commercialize it, ask what minimum performance it will guarantee in return for exclusive rights.

Urge the key decision maker to think in terms of minimum performance. Say: "If your engineering staff developed a new product idea, the company would require a certain minimum sales performance to keep that product in your product line. Regardless of who contributes the invention or what you believe the ultimate potential for this invention is, what minimum level of performance is necessary to keep the item in your product line?" This helps put the question in a context the decision maker can work with.

If your submission strategy allows, you might ask this same question of two or three competing companies in the marketplace. The various answers will further define the value of your invention and help you determine your licensing strategy.

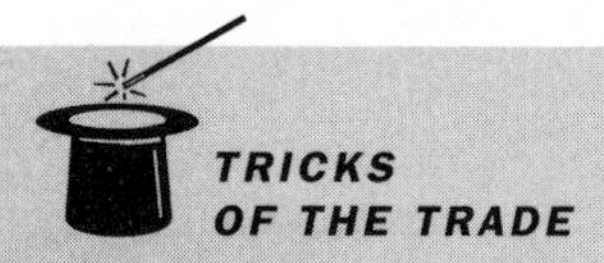

TRICKS OF THE TRADE

Up the Ante—If you're talking to several companies and your preferred company is giving you lower performance figures than the others, you may have some leverage. When a number of companies are attempting to gain marketshare in its industry, your preferred company may increase their bid to attract you away from the competition. This strategy is particularly useful when these companies are aware that you are in contact with their rivals. By now you know whether you are working in an industry where this is okay.

You would want to do this only if your technology is protected by a patent or if all the companies have signed a confidential disclosure agreement, or at least a proprietary agreement in which they agreed to honor your proprietary rights to your invention whether or not it is kept confidential.

This approach would not be appropriate when companies in your industry are particularly cautious about letting their rivals know about their product development plans. In certain areas of the medical and dental fields, for example, a company wants to be assured that it will be the only one to introduce a new product. It would adversely affect the company's reputation if it introduced a product that someone else has already offered.

The objective here is to determine what a company projects as the minimum performance for your invention over its life span. This information could reveal some of a company's limitations as a licensee. A company's distribution may be limited, or its ability to develop your product might be questionable. A wide spread between the minimum performance figures from different companies may indicate which company is best (or least) able to meet your invention's market potential (or that there's some wild guessing going on).

When the companies respond with relatively similar minimum performance figures, pay attention to whether the limits are dictated by the company's potential to perform or by your invention's potential. One way to determine this is to challenge the companies to offer higher minimum performance figures. If there's little movement, the limitation is likely your invention's potential.

QUALIFYING MARKET PROJECTIONS

When potential licensees assess the market potential for your invention, delve into the specific reasons for their projections. Earlier in this book we talked about certain phenomena that affect markets, for example, that safety does not sell, or that your item may be nonessential to most people, in fact may create a change they would resist. There are more instances of resistance to new products than there are of acceptance.

It is also possible that your invention is entering a crowded marketplace in which several other types of inventions have been introduced recently. Buyers might perceive your invention, although better, as just another product.

An example of this happened in the marketplace for sun visors. There was a period of rapid development for sun shading and sun visors in vehicles. Consumers have always been interested in improving shading in vehicles. Between 1989 and 1992, hundreds of inventions were introduced to improve sun shading and cooling in cars. From stick-on sun visors to suction cups to solar-powered cooling systems for cars, you name it, and there was a product developed for the consumer.

During the early period of development, many companies took on these new products. After two or three years, retail stores and consumers had had their fill of sunshade products. They wanted new products that offered something different. You might have developed the world's most advanced sunshade in 1993, but your invention would be relegated to the closet until the market became interested again.

The economy and the current competition within given markets can affect an invention's sales potential. We are talking about supply and demand here. You may be sure that everyone can use your product. However, market size is not a matter of how many people can use your product, but how many people will actually buy it.

Think about the products in your local supermarket or local hardware or automotive parts store. You could probably find a use for half of the items on the shelves. But as you know, neither you nor anyone else automatically purchases everything that you could use.

Consumer choices are dictated by many factors, including: your budget, the number of retail outlets in your area, the mix of products offered in those retail outlets, the relative prices of items you purchase, advertising, coupons and promotions, and your own personal preferences and values.

Sales promotions for your product will cover a large area, possibly all of the United States or even several countries. The bigger the market area, the more you must rely on other people. Your personal enthusiasm can't be conveyed to a nationwide or worldwide sales network. Your invention must sell itself as the market expands.

The Value of Wearing Someone Else's Shoes—It is important to understand the perspective of the person with whom you are negotiating. What is the real value of your offer from the company's perspective? Does your invention help to round out its product line? Will it give the company a chance to advance against a key competitor? Does it help the company maintain its market position? Will the worker's union like it? Might it save the company from a large capital equipment purchase? Or will it simply be its next cash cow?

It is also important to build a strong relationship with the key decision maker, whether in person or over the telephone. This relationship gives you an opportunity to talk about issues beyond the specific details of your invention and get into the company's values and philosophies. When intense license negotiations are stuck, revert back to what the company sees as your product's original value. Bring the discussion back to talk of value instead of negotiation. It is important to keep the key decision maker focused on visualizing this value as you go through this process.

The availability and price stability of the raw materials needed to make your product or technology will also affect the potential gross profit. Inventors do not normally have the resources and knowledge to weigh all of these factors. When potential licensees consider the projected sales for your invention and the ultimate value of your invention to the company, they must weigh all these factors. Therefore, when three or four companies respond with similar evaluations of your invention's potential, you can take this to the bank, so to speak. This is especially true if the results were concluded independently.

Foreign Licensing and Patenting

Your invention may have great potential in foreign markets, but receiving royalties from your invention in those markets can be challenging for several reasons:

- Foreign patents can be quite costly. It is not uncommon for the cost of foreign patenting to range from $20,000 to $100,000 (and up). In addition there are substantial fees for maintaining foreign patents.

- Accounting for royalties from foreign countries can be expensive and cumbersome to verify.
- Consumers in many foreign markets are loyal to products made in certain countries.

If your invention is produced in an undesirable country, you may not gain acceptance. Although we are entering the age of the global market and harmonization of patent laws, it can still be difficult to enter foreign markets.

If you want to pursue foreign markets, you can choose to license to a U.S. company that already has distribution in foreign markets. If your prospective licensees have distribution in foreign markets, ask them which foreign markets present opportunities for your invention and to what degree. This will be part of your overall licensing strategy.

The trick here is to get the U.S. company to agree to pay you royalties on its foreign sales. If your invention offers the U.S. company a favorable market advantage in the foreign market, it may be interested in a comprehensive deal that includes payment of royalties both in the United States and in select foreign countries.

Your foreign patent protection plays a role here. If you have U.S. patent protection but not foreign patent protection, this may hurt the company in the foreign markets. For example, let's say that a company agrees to pay you a 5 percent royalty for sales of your invention in Germany. The U.S. company would then introduce your invention to the German marketplace without patent protection in that market. A German company could recognize the opportunity and, in the absence of any patents, pursue its own market with your invention.

The German company has three advantages: it is operating in its own market, it doesn't have to absorb the cost of introducing the product since the U.S. company just did that for it, and it can offer the products for 5 percent less since it doesn't have to pay you a royalty. The U.S. company would have little incentive to enter the foreign market with your invention. You might have to negotiate a reduced royalty rate for foreign markets in the absence of foreign patent protection.

You may find a U.S. company that is interested enough in those foreign markets to pay for foreign patenting on your behalf. This may require you to assign your foreign patent rights to the company. The company would pay for your foreign patenting, and you would receive cash or a royalty according to the relative contribution of each party.

WORDS OF WISDOM

Even if your royalties from foreign markets are slim, the increased manufacturing production may help drive down your product's cost, thwarting competition and increasing U.S. sales.

Structuring a Licensing Deal

By now you have a good idea as to which companies are seriously interested in licensing your invention and to what degree. Learning from the information you have gathered thus far will help you develop the best possible deal. Consider all the various markets for your invention, their sizes, their geographic area, and which potential licensees service those markets. Now, let's look at some ways a licensing deal may be structured.

EXCLUSIVE

An exclusive license is one of the more common license agreements between an inventor and a company. The company agrees to manufacture and market your invention nationally or internationally and pay you a royalty or other remuneration in return. Minimum performance standards should be factored into exclusive licenses. Exclusive licenses can have time limits of three to five years, for example, which give the company a head start on production and allow it to capture profits to pay for the startup costs. By the time the product is established, the license would be converted to non-exclusive, allowing other companies to jump in the mix.

> **Last Chance to Bail**
>
> If you are still considering whether or not to pursue your own entrepreneurial effort, now is your last chance to walk away from a licensing deal. You know the competition's limitations and predispositions, you have made no commitments to them, and because of the agreement you had them sign, they can't compete against you with a similar product. If you're going to walk away, do it now. Otherwise, strap yourself in for the ride.

NON-EXCLUSIVE

Non-exclusive licenses are particularly appropriate when your markets are extremely fragmented and no single company commands more than a majority of the market and distribution share. One drawback here is that the competition among your non-exclusive licensees may bring down the price of your invention. If you foresee this happening, you may want to base your royalties on each manufactured unit as opposed to the sale price of each unit. An option is to license to only the top three or four companies so they don't bump into each other and crowd the market.

SEPARATE MARKETS

Perhaps you find different and distinct markets for your inventions. For example, there may be a mail-order market and a market through retail stores. Let's assume that one company is appropriate for manufacturing and marketing your invention to the mail-order business and another company would be appropriate for manufacturing your invention in a different form and marketing it to the retail stores. You may want to license to each company exclusively in its respective markets. Neither company would be able to cross over into the other company's market. This way you can effectively capture both markets. This same scenario may apply when your invention has applications on both a consumer level and an industrial level. You would segment those two markets in a similar manner.

DIVIDED TERRITORY

You may find that one company is dominant in the markets west of the Mississippi and another is dominant east of the Mississippi. There may be other regional variations. You may decide to license to companies exclusively and individually by territory. This scenario would apply more for service-oriented inventions or those for which long-haul shipping is an issue. A drawback of this is that if each company had to produce your invention, none of the companies would realize the full efficiency of mass production. When dealing with services and methodologies, however, these functions tend to be more geographically oriented and are not affected as much by centralized manufacturing.

There are as many commercialization strategies as there are inventions.

DELEGATED FUNCTION

In this scenario, Company A is in a position to effectively market your invention but does not have the resources to manufacturer it. Company B has the resources to manufacture your invention but not market it. Therefore, you may award Company A the exclusive rights to market your invention and Company B the exclusive rights to manufacture it. However, the *Doctrine of Exhaustion of Rights* may prohibit you from receiving royalties from both the manufacturer and the marketing company for the same product. This law can limit you from receiving royalties from more than one level of the distribution channel for the same invention. Therefore you would need to evaluate which company you want to receive royalties from.

The royalty rate from a manufacturer may need to be twice the rate received from the marketing company to total the same amount of money. The marketing company may sell your product for twice the price as the manufacturer. Therefore a 5 percent royalty from the marketing company would yield twice as much money as the same rate from the manufacturer. If you can influence the manufacturing company to pay you a higher rate, you gain the advantage of receiving your money sooner since you don't have to wait for your product to go through the channels of distribution to the end user. Also look at the solvency of both companies to see whether one is in a better financial position.

LEAST COST MANUFACTURING

This strategy is particularly beneficial when dealing with several non-exclusive licensees. It is not uncommon for several companies to be interested in marketing your product without a license and without paying royalties. However they may not want to manufacture your product and would prefer to purchase the product from the manufacturing company that is working with you. In this case you would choose one manufacturer to exclusively produce your invention, which would then be distributed to those various marketing companies that would distribute and sell your invention under license.

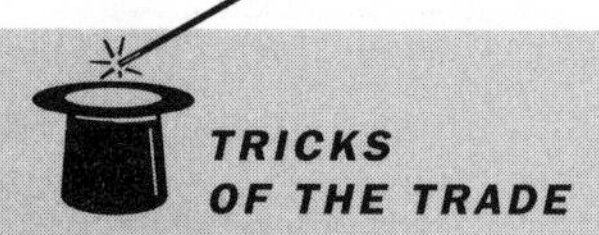

The Best Deal in Town—Consolidating orders and obtaining the least cost manufacturing is one way to make money from an invention that is not patented. If you find a low-cost manufacturer that can provide you with volume discounts and a fragmented market in which there are several companies in a position to market your product, you can offer them your product at a lower price than they could get on their own. This might be incentive for them to handle your invention exclusively through you, with or without patent protection.

You may be able to persuade the various marketing companies to offer you a purchase order or, even better, an irrevocable letter of credit that holds their money in escrow. When you deliver your product to them at the specified price, and it meets the predetermined specifications, their monies will automatically be transferred to you, or to your manufacturer's account.

With several of these purchase orders or letters of credit in hand, you can receive higher volume discounts than any single marketing company. You can therefore get your product to the marketing companies for a lower price than any of them could on their own. You may be in a position to act as broker between the manufacturing and marketing companies and to earn a profit between the two. This would also give you greater control of your product's sales volume and your income.

Or you could license your manufacturing company to pay you a royalty on its sales. This may remove you one step further from the loop, which could be important if you want to reduce your exposure to product (legal) liability.

Joint Venture Opportunities

Several of the scenarios already described are essentially types of joint ventures. Joint ventures can be established with other inventors too. You may have found that your invention needs to be modified before it is marketable and another inventor could help you. Or there may be other inventors and patent holders who have developed technologies that complement your patent or trade secrets. If so, it would be prudent to seek out those other inventors and learn how successfully they have commercially exploited their inventions. They may be faced with the same commercialization tribulations you are encountering.

They may have not researched the marketplace as thoroughly as you have through your interviews. You may be in a position to suggest that combining your expertise in a joint venture would more effectively accomplish both your goals. Since most patents sit dormant and most inventors do not have the fortitude to go through the laborious process outlined in this book, you may be offering them the deal they need to keep going. If you have an invention you haven't patented (or even an invention that needs slight modifications) or know-how that you could combine with another inventor's patent to create a better product, a joint venture could pave your way to approaching potential licensees immediately, rather than waiting for your own patent to issue or without having to make costly modifications.

TRICKS OF THE TRADE

Government Freebies—A variety of technologies that were initially developed by the U.S. government or military are now available for the private sector to commercialize—many of them free of charge. Leverage this resource by co-licensing a government patent and adding it to your collection of intellectual property. Contact the specific agency (NASA, the Department of Energy, National Laboratories, all branches of the military, and so on) that develops topics of interest to you.

One way to find complementary inventors is through a global search for recent patents in your subclass. There are also clipping services for retrieving copies of patents specific to a particular subclass. Or you could order a search of all the patents in your subclass at the USPTO. These are also listed on the Internet at www.USPTO.gov and at patent depository libraries.

CO-LICENSING

When two or more inventors combine their patented technologies and offer them for commercialization, this is called co-licensing and sometimes referred to as cross licensing. Even though these terms are sometimes used interchangeably, a cross license specifically refers to an arrangement whereby you license your technology to another party, and that party licenses their technology back to you; therefore, the inventions cross between you. For example, if you utilize Company T's patented technology to enhance your patented technology and then license the finished product to Company T for commercialization, you are considered cross-licensing partners.

If you co-license, be sure you have written contractual agreements with the other inventor. A patent attorney who is experienced in licensing can help you create an agreement of this nature. A co-license agreement can be written so that it does not unduly bind you if you later decide to operate independently of the other inventor. You are free to structure a co-license however you deem appropriate. For example, you may want to have a provision that either inventor would be free to license and manufacture independent of the other inventor based on certain criteria that would be favorable to both of you.

Fear Not the Foreigners—Some inventors hesitate to do business with foreign manufacturers because they may sell your invention to other countries without your knowledge using, for instance, the molds that you paid for. Although this seems like a grave injustice (and it is!), it can actually benefit you in the long run. Why? Because when your foreign supplier increases its volume (to fulfill all those clandestine orders), it can lower your product costs and help keep you competitive in your home market. You'll never be able to completely control sales in foreign countries, so why not use it to your advantage on the pricing end? That's one way to make a silk purse from a sow's ear.

This list of licensing variations is meant to help you expand your thinking about licensing possibilities. Licenses are not limited to the specific forms listed here. There are as many ways to structure license agreements as there are people.

Life after Rejection

When you believe that you have contacted all the potential licensees for your invention, and they have all rejected your invention for similar reasons, take a hint. You may have the right idea at the wrong time. Although rejection is a hard thing to accept, a measured approach to potential licensees will have saved you the frustration of spending exorbitant amounts of time and money on patenting, prototyping, or marketing a futile endeavor.

APPLY FOR A GRANT

If your rejected invention would make a significant contribution to humanity, there still may be hope. The federal government and other agencies have programs, grants, and resources to help inventors further develop their inventions, even if they aren't patented. The Resources section of this book has a comprehensive list of programs that are particularly appropriate for independent inventors.

GO ABROAD

Consider taking your invention outside the United States. Foreign patents can help you take advantage of this. Or you might find a foreign company that is willing to do the initial manufacturing and marketing in its country without any patent, but saving the U.S. rights for you should you obtain a U.S. patent. Although you may not earn royalties from sales in a foreign country, the results of the marketing in that country may encourage a U.S. company to take on your invention once it has been proven overseas. This would be analogous to a free test market. One example is the hemp business. Growing hemp is currently illegal in the United States, even though products made from its fibers were used successfully in WWII (in parachutes and other applications). These days the fibers, which must be imported, are used in hats, clothes, and

There is life after rejection.

other durable goods. Supporters hope positive experiences growing hemp in other countries will generate acceptance in the United States and open new opportunities for patented processes in this field—similar to the way in which the introduction and positive sales of my wedge mirror in Canada opened the door for U.S. sales.

START A BUSINESS

Another avenue you can take is to manufacture and market your invention on your own. This requires personal commitment, entrepreneurial effort, intestinal fortitude, and financial resources. You may be able to start out small with a cottage industry and pursue alternative marketing avenues such as mail-order, multilevel marketing, public relations blitzes, and the Internet. These methods reach the consumer directly. After you have established yourself in the marketplace and proven that your invention does indeed have perceived value, you may be in a position to license or sell to the major corporations that weren't previously interested.

To hell with circumstances; I create opportunities.

–BRUCE LEE

Inventors who have succeeded with these alternative methods of getting to market did not take no for an answer. Remember, though, that for each inventor who has succeeded there are at least a hundred others who have failed. As we said earlier, even some progressive corporations, with all of their resources, succeed with only three out of every ten products they introduce. Think of where that puts your odds. This is not to discourage you, but rather to forewarn you of the risks you face.

If you choose to go the entrepreneurial route, you need to create a business plan. And since you have the results of all your great market research, you're halfway there. *Your First Business Plan: A Simple Question and Answer Format Designed to Help You Write Your Own Plan,* by Joseph Covello and Brian J. Hazelgren (Sourcebooks, 2005) and the U.S. Small Business Administration website (www.sba.gov/smallbusinessplanner/index.html) have easy-to-follow, fill-in-the-blanks formats. There are dozens of other books and resources like this to assist you. See the Resources section for more info.

Acquiring investment capital for your product may not be as tough as you imagine. Investors frequently complain that they cannot find quality projects with savvy management, a marketable concept, and a good business plan. If you have these three elements, you'll find the money!

Most major cities have entrepreneur and investment clubs that can usually be accessed through the local business library and a U.S. Small Business Administration (SBA) office. The SBA has several programs for budding entrepreneurs, and one of my favorites is SCORE. This is a free consulting service where retired business executives volunteer to give advice to neophytes. It's nice to sit down with someone who has been there, done that. However, be careful. I've used this service and once got matched with someone who just liked to hear himself talk, and his experience was not appropriate to my situation. Make sure the program supervisor matches you with the right person by communicating your goals clearly.

POST-REJECTION PATENT CONSIDERATIONS

Even if you receive universal and bonafide rejections from everyone in the industry, it still may be appropriate to consider patenting. This is true when the rejections indicate that there is merit to your invention, though the market is not ready for your brainchild at this time.

A good question to ask key decision makers is, if not now, when? Acceptance of your invention may be only five to ten years away. You may need to proceed with the patent process within the next year to avoid being time-barred from receiving a patent.

From a timing standpoint, you are gambling the cost of applying for a patent and paying maintenance fees on the patent against the hope that the perceived market value for your invention will come of age in time for you to financially benefit. You may want to instruct your patent attorney or agent to take every opportunity to extend the amount of time it would take to receive the patent so certain fees are not due as soon. This procedure may or may not have an adverse effect on foreign patent filings. Discuss that matter with your patent attorney or agent. Another way to extend the life of your patent application is through continuations and continuations-in-part. Consult your patent attorney or agent about this also.

THINGS TO REMEMBER AND CONSIDER

- Develop your submission strategy carefully.
- Decide what to submit and whom to submit to.
- Determine ahead of time what you want to learn from key people in your industry.
- Before moving forward, recognize and consider all the factors that will go into making your project successful.
- Have an exit strategy, and be willing to retreat with grace.
- Be ready to identify other opportunities.

CASE STUDY

The Picture Perfect Deal? (Part 2)

We rejoin the story from the last chapter as I was preparing to meet with the president, the engineer, and my client, the inventor, in Chicago. Our goal was to test the prototype and review the terms and conditions of a possible license deal. On the way there, I went over the questions I wanted to ask and the issues I wanted to be sure to cover. I planned to ask the first set of questions during our tour of the facility.

SALES PER PRODUCT

I wanted to get an idea of the usual size of the company's production runs. For example, did it have a minimum production run of five hundred units per item? What portion of its product line exceeded this, and by how much? Was it in continuous production of all items, or was production staggered; if so, how often? What production process did the company subcontract out of the factory? What type of backlog did it have? What was its history on backlogs? Answers to these questions would give me a feel for the type of volume the company was used to, and help me understand how my client's invention would fit into the company's production process.

RAMP-UP TIME

Once an invention reached the prototype stage (as this invention had), how much time did the company anticipate it would take to actually get production units out the door? What had been its past experience with ramp-up time for similar types of products? How many units could it produce without making major changes in the current production line? What resources and investment would be needed to increase production? Were there any other new products under consideration that would utilize the same resources?

DELAY PROBLEMS

What types of problems created the biggest headaches for the company, and how did this affect the delivery of products? Most production lines have an Achilles' heel where something breaks down frequently, causing bottlenecks. What was this company's Achilles' heel, and how would it affect shipment to its customers? How often had it happened? If there were delays in production, how long did they last?

NEW PRODUCTS

Of the new products the company had introduced in the last couple of years, which, if any, had been from outside sources such as inventors or licensed from other companies? What had been the company's experience with working with inventors or engineers from the outside? What type of success had it experienced with these product introductions?

SHIPPING AND RETURN

What was the company's return procedure? What percentage of its products were returned, and for what reason? What were the volume and rate of return for various products? What had been the company's experience with shipping overseas? How much and how often did it do so?

QUALITY CONTROL

What was the company's procedure for quality control inspections, and which of its products underwent inspections? What was its rejection rate and history in this respect?

PARTS SUPPLY

From what sources did the company receive its raw materials and parts? What type of complications had it experienced from its suppliers? Had it recently changed suppliers, or did it have a long-standing relationship with most of them?

EMPLOYEES

Were the company's employees unionized, and had there been recent changes or were there anticipated changes in this respect? What was the average longevity of the work force? Did it offer long-term profit sharing, or did it hire seasonal workers? The primary reason for the last couple of questions was to get a feel for what type of company this was and what its general attitude toward people was. One of the best ways to understand how you may be treated as a licensor is to look at management's and ownership's relationships with the people with whom they do business. This would help me understand their company values and know what to expect in a business relationship with them.

I knew other interesting tidbits of information would come out of the factory tour. Next we would witness the engineering prototype test. The conversation at the negotiation meeting would then take on a very different flavor. There we would turn our attention to the terms and conditions of a potential deal. I wanted to cover the following topics in the meeting, for the named reasons and strategies:

THE TEST

The first thing I wanted to do in the negotiation meeting was to ask the company representatives what they thought about the engineering prototype test. I wanted to do this even if I already had an idea about how they felt. If the test was very positive, I wanted to reinforce this. If the test provided questionable results, I wanted to discuss any reservations they had, what hurdles needed to be overcome, and how they felt about overcoming those hurdles. I have come to this point in many negotiations only to realize that the company officials saw obstacles they had not yet revealed.

Now is the time I would lay all the cards on the table. If people had reservations, I would address them now. If there were significant problems, we might have to adjourn and reconvene after the problems were solved. A contingency plan might assume that the problems will be solved. In this event, the parties might agree to go ahead and outline the terms and conditions of a potential agreement that will be in place when it is time to proceed. This inventor had already tested the same prototype that the company was testing, with positive results. So we assumed there would not be any earth-shattering revelations that would preclude hashing out the terms and conditions of a license contract or buyout.

THE PRODUCT

My primary objective was to encourage the company officials to visualize the invention as a new item in their product line. I wanted to draw them into the picture with me. I would describe the variations of the invention, the number of different items that might be formed around the patent or technology, the different sizes and models, and so on. This could also start the creative juices regarding additional products that may be generated from the invention.

POSITIONING

How would sales of the invention affect the company's position in the marketplace? How would customers view the company as a result of introducing this invention? Did the company want to gain or maintain a position of being innovative, and how important was this to the company? The answers to these questions set the stage for the rest of the negotiation and established additional value.

EFFORT

What type of effort did the company anticipate putting into this item? Where did it see the market? Whom would it sell the product to? What types of resources—people, equipment, engineering, technology, and finances—would be needed to launch this product? How would the sales force present the product? What type of advertising and promotional program would be needed? (This may seem redundant, but it can't hurt to reinforce value and keep the company focused on the product.)

TERRITORY

In what areas of the world could the company effectively distribute this new product? How effective did it think it would be? Where would it face its stiffest competition? This was important even if the invention was the only one of its kind, because the company's ability to distribute a product in a given geographic territory may depend on its ability to gain access to that market in general. If an invention is one of five hundred items the company offers, it will have little impact on whether it enters or expands into a given market segment. On the other hand, if an invention would make a significant difference in its expansion plan, this adds to the invention's worth.

SCOPE OF LICENSE

What exactly did this company want to license or purchase? Did the benefits lie in the patent rights and/or improvements of modifications thereof? Did the inventor have trade secrets, know-how, and/or other attributes of interest to the company? To what degree did the company value these additional things?

TIMETABLE

Taking into consideration the ramp-up time, and adding the extra time it might take to introduce the product to the marketplace, when did the company anticipate any sales of substance? How long would it take to reach certain milestones? For example, if the company intended to start with one item and then expand to two or three, when might this expansion be anticipated? Companies hate to nail down specific timeframes; however, I wanted to ascertain at least a range of possibilities—a minimum and maximum period.

PRICING

Which items would sell for what price? Would the company maintain a higher sale price to begin with and then lower it in time? Would it start with a low introductory price and then increase it, and if so, when?

ASSIGN OR LICENSE?

By now, the key decision makers in the company would have thoroughly visualized the invention's applications and would understand what it could mean for their company. Although they had already expressed interest in licensing rather than an outright purchase, the inventor was interested in an outright purchase if possible. Therefore, I would ask whether the company was interested in purchasing the patent right (having the patent right assigned to them) or license, whereby it could spread out its cash flow over a period of time.

It is not uncommon for a savvy company executive to suggest that an invention be assigned to the company and that it would make payments to the inventor over a period of time rather than making an outright purchase. The company might offer an exchange of stock. I would advise against assigning the patent unless there was a substantial initial lump sum payment. The company might choose to lower the payment somewhat in exchange for scaled-down royalty payments over time, and this might be acceptable.

The reason I wanted to see a substantial lump sum payment for assignment of the patent was because once the patent was assigned to the company, the inventor would lose ownership and most of the control of it. Remember the old saying, possession is nine-tenths of the law? If the company were to renege on its commitment to maintain ongoing payments, it could mean a lengthy and expensive procedure for the inventor to repossess his patent.

I expected them to choose the license route because the company had already shown that it wanted to conserve cash flow by the way it had handled joint development with the inventor. Nonetheless, while the company was visualizing all of the great potential for this invention, it was worth a shot to ask. If it said no, we would simply proceed to hash out the terms of a potential license.

ROYALTY RATE

I wanted to set the royalty rate now, after we had determined the invention's great value, and that it would indeed be a license deal and not an assignment. I wanted to determine the royalty rate before going into further details and the necessary limitations of the license. I also wanted to discuss this rate independent of any consultation fees that might be paid to the inventor for ongoing assistance.

It is not uncommon for a company to assume that an inventor should be expected to provide ongoing assistance in exchange for its paying royalties. We needed to remember that the basis for royalties is the hard work the inventor had already put in over a long period of time. Thus far, the inventor had taken all the risk, in both time and money. Putting it this way would also remind the company of the value of what it was paying for. It also would suggest that the inventor might earn extra fees for ongoing consultation.

The first thing to determine is what the royalties would be based on, e.g., gross sales versus net sales or number of items sold. This would primarily be a function of what the company had been used to doing in the past or what was practical for this type of item. It also depended on the company's accounting system and the easiest way to track the invention and differentiate it from other items being sold. This being said, I would want to talk about royalty rates. I hoped we could leverage the company's own enthusiastic indication of how great this invention would be for the company.

I would insist that the company recommend a royalty range, and then I would attempt to negotiate something no less than the highest end of the range. It is usually effective to counter with a figure high enough to make the decision makers stutter a little. Once I make a counter offer, I do not speak until they have. Their response to the counter offer would set the stage for the additional negotiations. Their response would indicate how far they were willing to move and what they saw as the areas of resistance.

I remember the old saying, "After you have made your offer, the next person to speak loses." My experience is that this is true. When you make an offer and then add some rhetoric out of nervousness because you are afraid they might reject it, you have effectively taken away any possibility for them to immediately accept your offer, or at least come close. I once waited in silence for a response to an offer literally for three minutes. When they still hadn't responded, I asked if they were still on the phone, then retreated back to silence. When they finally did respond, it was positive.

Establishing a royalty rate would provide a baseline from which we could potentially add and subtract as we negotiated other aspects of the total payment plan.

NON-EXCLUSIVE VERSUS EXCLUSIVE AND MINIMUM PERFORMANCE

I believed that this company would want an exclusive license arrangement, and as with all exclusive licenses, I would insist on a minimum sales performance expectation. I would also need to determine what would happen if the company failed to meet the minimum performance. Two common options are for the company to either lose its exclusive license while maintaining an opportunity to manufacture and sell the product, albeit with licensed competitors, or for its license to be rescinded.

I like to allow the company to set forth the initial minimum sales performance expectation. If the decision makers stuttered here, I would ask them at what point they would likely yank the product from their product line. The company had thresholds at which it would not continue to offer a product if the sales were too low to justify it. The main thing I would look for here was that the minimum performance they suggested now was not too far out of line with the one they gave earlier in the conversation. This is another good reason for visualizing the sales potential before introducing the minimum performance aspect of the license negotiation.

Another aspect of minimum performance is timing. If a market is sporadic, it may be unrealistic to tie minimum performance into a single calendar quarter. The performance figure may need to be averaged over a longer period, such as six months or a year. On the other hand, we don't want to go too far out; if a long period of time elapses and the company fails to meet the minimum performance standards, it delays our ability to re-license to another company. The number of calendar quarters that the minimum performance is based on would be a combined function of the inventor's tolerance for a longer time period and the company's tolerance for a shorter time period.

INITIAL PAYMENT

Initial payments are looked at in two different ways: initial payments that are actually advances to be deducted from future royalties, and initial payments calculated independently from royalties or consultation fees. I would assume the latter definition of initial payment would be used unless and until the company assumed the former definition.

On a pragmatic level I had discussed with the inventor the advantages of steering more toward initial payments at the sacrifice of ongoing royalties, versus lower initial payments in favor of maintaining higher ongoing royalties. A bird in hand is worth two in the bush, especially with the circumstances surrounding this invention. If this licensee didn't succeed, who else effectively could? There were limitations with this project, but we wouldn't show our sweat.

It might be hard for the inventor to make decisions on the spot during the negotiations. This would be an excellent time to suggest a short break so the inventor and I could consult regarding the choices facing us.

The inventor might have a lot to consider about initial payments: how much he had invested in this project weighed against how long it would be until he could receive substantial royalties. The final deal would be a compromise between what the inventor wanted and what the company would offer. I, for one, certainly hoped that the parties would give and take. It makes my day when deals like this don't go up in smoke over some petty issue on which neither party is willing to compromise.

ADDITIONAL PATENTS

Who would pay for the maintenance fees on the patents? Who would pay for the application of additional improvement patents when developed primarily by the inventor for use with the existing invention? Who would pay to file international patent applications, and what would the patent strategy be for obtaining proprietary protection overseas? I would suggest that if the company was interested in an exclusive license, it should bear the burden of these costs, or at least a significant portion. After all, it would do this if it developed a technology in-house for its own exclusive use. Incorporating the inventor's patents that were already in existence would save the company time in entering the marketplace quickly. The company had not had to assume any risk at this point. Whenever I am stuck in negotiation, I always go back to the value. I keep everyone's eye on the prize. I have them reflect on those positive visualizations.

INFRINGEMENT

Who would pursue infringers? I would take the same stance here that I did regarding maintenance fees. As long as the company maintained exclusivity, it should take the lead in pursuing infringers. Whichever party pursued infringers, the other party should have an opportunity to pick up on any relevant lawsuits if the first party fails to do so. This provides an insurance policy of sorts on the patent protection. In the event that both parties contribute toward an infringement action, the parties would share in any proceeds proportionate to the relative amount that each party paid into the litigation.

This is one of those items on which it would be easy to be flexible since there was little chance of significant litigation on this invention, owing to its niche market status; the sales potential simply would not justify it. Nonetheless, if I offered the company an opportunity to choose the terms of this aspect of the license contract, it would show good faith and reasonableness and afford a psychological opportunity for me to stand firm on other terms of the contract.

ACCOUNTABILITY

How in the world would the inventor know if the licensee was being honest and forthright with its royalty payments? Well, to begin with, if the company paid a minimum performance amount, it didn't have much of a choice. I could suggest some wording that would afford the inventor the opportunity to inspect the company's records. It was unlikely that this would ever happen. Fortunately, it is easy to keep abreast of the sales in this field by querying distributors and seeing whether the extrapolation of their sales closely matches the royalties being paid. It is kind of a bottom line way of being accountable. Therefore, I would suggest flexibility on this term.

SPECIFICATIONS AND QUALITY CONTROL

I would suggest that as long as the company was maintaining a product that met the overall objective as outlined in the patents, this was adequate quality control. The company would probably want to maintain control over the exact specifications as well as responsibility for the quality control. This would give the company control over the process and also place product liability on the company to maintain a quality product. If the inventor assumed any of these functions and there was a liability lawsuit, the inventor could be dragged into it since he would have had more of a role in the development of the finished product.

I believed it would be more palatable for the company and better for the inventor to have the company ascertain the product's exact specifications and maintain quality control. The inventor, being something of a consumer advocate, could make certain that the company designed the safety and performance elements of the invention as he intended. This would help to put the inventor on the side of the consumer.

I further believed that the company should remain open to considering input from the inventor for improvements and modifications. This would set the stage for a possible consultation arrangement between the inventor and the company.

This would be an excellent opportunity to sneak in a term of the contract that specified an hourly rate for the company to pay the inventor if the company sought or used the inventor's advice on an ongoing basis. If the company did not ask for advice, it wouldn't cost the company anything. If the company was reluctant, I might suggest a compromise whereby we put this consultant term in place, and the company would automatically receive a certain number of consultation hours per calendar quarter at no charge. Any consultation over and above this would be billed according to the terms of this agreement. Sounds reasonable, doesn't it?

INSPECTION

I would recommend that the inventor be afforded the right to at least an annual inspection of the company's facility. I was not sure whether this was important, but it would give him an excuse to do business in Chicago occasionally. This was one of the best ways for the inventor to make sure his products were being manufactured within the scope of his original intentions. The personal touch doesn't hurt either.

I would suggest that the company send the inventor a free sample or two of the invention every so often. This would not only help from the inspection standpoint, it also would be nice for the inventor to be able to show off production units of his invention without having to pay for them. I would invite the company to make an offer in this respect.

During our day of tour, test, and meeting, I was able to discover most of the information that I set out to get. The engineering prototype test went better than expected. The negotiations, overall, went very well. The company suggested licensing with a royalty of around 15 percent (which I considered to be quite generous—twice what we would have accepted) plus an initial payment. It is important to note that the royalty rate was the company's suggestion, thus proving the value of encouraging the company to make an initial offer, then waiting, in silence, until the cows come home for the response. I wrapped up the meeting by nailing down a timetable to bring the discussion to a final conclusion in the form of a license contract. I offered to generate a proposal containing all the salient terms we had discussed as a preliminary to a formal licensing agreement, which I would also prepare. As the company does not typically retain a patent attorney and was somewhat "attorney-averse," the decision makers were relieved that I would take the lead in this regard. To reach finalization, we agreed to keep an open dialogue regarding initial payment and minimum performance, which were points we still needed to agree on.

As my client and I hoped to finalize the deal within sixty days of the meeting, I had the proposal to the company within a month. It looked like we were ready to strike a deal. All that was left to do was to follow up with the company president and work out the details necessary to come to a final agreement.

Three months after the big meeting, I was finally able to make telephone contact with the president to discuss the proposal. The president said that the company had decided to pass on this deal. Surprised? Well, I was certainly surprised. The first thing I asked was, "Why?" My negotiations with this company had been moving along in a positive manner for many months. The company had gone to the expense of building prototypes and product testing and had even agreed to pay the inventor advance money to cover patent maintenance fees.

The president said that the potential market for this product was too small, compared to the effort the company would need to put into developing its market position with regard to an electrical unit versus gas to be worthwhile. The company was considering three other projects, and its resources were limited. It simply came down to weighing how much it needed to put into a given project versus how much it would receive back, and my client's invention ranked fourth in a list of four. Something had to give. It wasn't lack of a market for the invention; there certainly was a market. It wasn't that the invention was unfeasible. It certainly was feasible.

But because it was in a niche market, the potential sales were not enough to excite this major national manufacturer. Does this mean the end of the road for the inventor? Certainly not. It means that we worked diligently for more than a year and tried to put a deal together with the best potential licensee we could find, and for reasons beyond our control, it didn't work. There are other companies on our B list that have a lesser presence in the marketplace. Since their expectations aren't as high, we hope they will be satisfied with lower sales than were necessary for the first manufacturer we chose.

HINDSIGHT LESSONS

If we had to do it all over again, would we do it the same way? Probably yes. There are several books that suggest that you should always go to the B list of companies instead of the major companies for licensing because of the smaller companies' greater flexibility and lower expectations. This may be true in some instances, but how do you know what will happen to you? There is no magical way of prequalifying companies based on their size. This is particularly true when major companies downsize and divide into smaller units, thereby operating more like a small company.

If we had gone to the secondary manufacturers first and received a very positive response, we would have always wondered whether we should have gone for the industry's major manufacturer instead. And had we communicated with the B list of companies during the same period of time that we were negotiating and doing product testing with our main potential manufacturer, we would have alienated our primary company and would not have received the advance fees we acquired to build a prototype. Such is the nature of this high-risk business. Everything can look great, yet ultimately not work out.

Has the inventor been successful? From an engineering standpoint, certainly yes. He has produced a product that does what he said it would, and it meets an unfulfilled need in the marketplace. But he just as certainly hasn't profited from his invention. In fact, he still has nearly $30,000 invested in it. This is a good example of why you should not mortgage your home or go too far in debt for your projects.

This story is meant to be a reality check, and it demonstrates the way inventions go more times than not. In retrospect one might suggest that the inventor should have formed a company and started producing units since there is evidence of some sales potential. When the potential profit margin and sales volume are high enough, this can be feasible.

However, remember that we were not anticipating millions of sales per year, rather a few thousand at best. We were also looking at a substantial investment to tool the unit, along with additional development time. This is hard to justify for a single item. If the inventor had an entire line of products, or if there were existing products on the market that were complementary, it could have increased his chance of success. After having said that licensing is a high-risk game, increase that risk ten-fold when considering starting a whole new business around your invention.

On a positive note, this story provides good examples of things to do and consider in the process of licensing your invention. You can do everything right with half the effort (as shown in the "Smooth Success" case study in Chapter 4) and reach a positive conclusion. Or, as in this story, you can do everything right with twice the effort and meet with failure. As you can see from these case studies, it's hard to predict the outcome even when you get to final negotiations.

A life spent making mistakes is not only more honorable but more useful than a life spent doing nothing.

—GEORGE BERNARD SHAW

CHAPTER SIX

Brass Tacks

Negotiating Your Compensation and Other Contract Terms

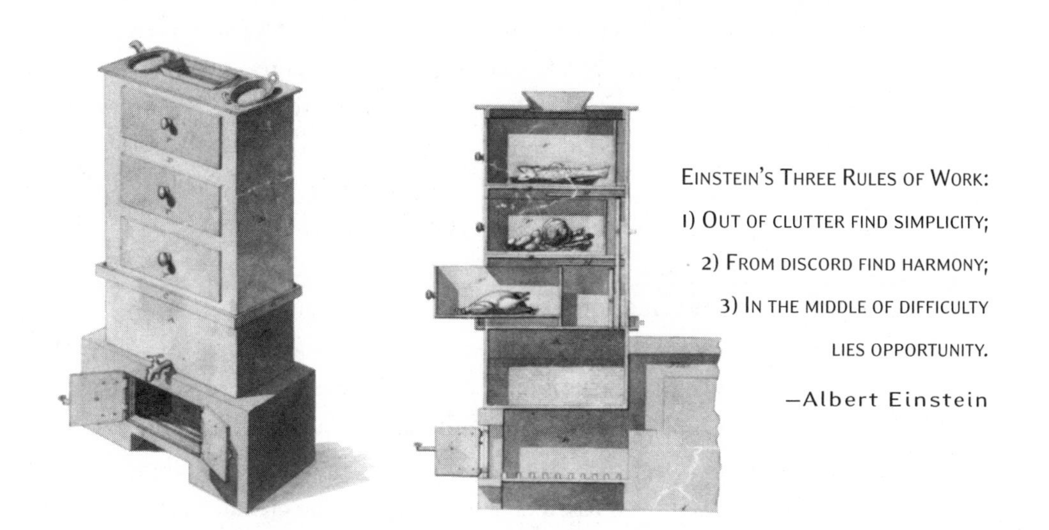

EINSTEIN'S THREE RULES OF WORK:
1) OUT OF CLUTTER FIND SIMPLICITY;
2) FROM DISCORD FIND HARMONY;
3) IN THE MIDDLE OF DIFFICULTY LIES OPPORTUNITY.

—Albert Einstein

In the last chapter, you learned about the market value for your invention, decided which company(s) you want to work with, and explored various ways to structure a licensing deal. Once you and a potential licensee come to agreement about the market potential for your invention, it is time to start negotiating compensation.

What can you expect in terms of royalties and advance payments? A lot depends on what you are bringing to the table versus what the company provides. Say you come forth with an unpatented idea, and a company is going to invest $2 million to develop it. Regardless of the potential for your invention, your bargaining power is not nearly as good as if you had a patented item, with all the engineering bugs ironed out, and had passed a test market with flying colors.

The main factors that affect your remuneration include:

- The market and profit potential, or with industrial inventions, the total savings to the company
- The degree to which your invention helps the company position itself in the overall market

Steam kitchen by John Bouis (U.S. Patent No. 1632X), Baltimore, Maryland, 1812.

- Your expenditures to date and any additional anticipated investment
- The company's anticipated investment
- The company's cash position
- Other valuable assets you may be presenting to the company

Royalties

When determining how much money you will receive for your invention, it all comes down to value. What is your invention worth, to whom, and to what degree? Royalties vary widely. Many companies pay a 5 percent royalty to inventors with little or no advance. Patent attorneys and licensing professionals use formulas that calculate projected sales and profits, and the inventor receives a proportion of that. More often than not, however, royalties are based on raw negotiation with the licensor (you) trying to get as close as possible to what the market (licensee) will bear.

WHAT ARE ROYALTIES BASED ON?

Sometimes companies base royalties on an item's profitability. This is a touchy area because you are at the mercy of the potential licensee to take appropriate action to make a profit. In some cases, especially with expensive items that will have a relatively low sales volume, a per-piece royalty may be appropriate. In general, royalties are based on a percentage of the manufacturer's net selling price. The manufacturer has no control over what the retailers sell the item for and prefer to commit to the income they can account for.

A lot can come down to raw negotiations.

Although you may start your negotiations at the gross sales price, most companies will work you down to a net sales figure. The gross price represents all monies received for a product. The net price is the gross price less deductions for things like returns, trade discounts, and shipping. It's reasonable that a company wouldn't want to pay royalties on UPS delivery charges or a product that was returned.

Sometimes the company will want to deduct sales commissions of 5 to 10 percent of the gross price. If you can get a 5 percent royalty on 95 percent of the manufacturer's selling price, or gross, consider yourself lucky. I have negotiated royalties as high as 12 percent gross for an inventor and have obtained as much as $25,000 up-front money just for the option to license an invention. I have also agreed to a settlement of a whopping $100 up front. In another case, 14 percent was deducted from the gross figure before the royalty percentage was calculated. Every situation is unique.

A company that is savvy with licensing negotiations will declare that the more money it pays the inventor up front, the less it will have available to promote your invention. This is a hard point to argue against, particularly if you are interested in the long-range commercial success of your invention.

Inventors often have already incurred substantial expenses for patenting, prototyping, and research and need to be reimbursed as soon as possible. Therefore, you can argue that the potential licensee should at least reimburse you for the out-of-pocket expenses that the company would normally have paid for a project developed internally. The company may agree to reimburse you for such initial expenses, but it may want to make it an advance payment against future royalties.

OTHER TYPES OF REMUNERATION

There are other types of remuneration than the almighty dollar. For example, your knowledge may be seen as a greater contribution than your actual invention. You may be offered a job or consultant fees to help develop technology in your area of expertise.

Or you may find that the potential licensee produces something you want. Let's say you invented a part for an engine and wanted to do more research and development on engine parts. You might strike a deal to receive engines as royalties to help with your ongoing development.

Remember the personal dimension as you negotiate.

Some Thoughts on Negotiating

Negotiations tend to be more effective when done in person or at least over the telephone. Many a deal has been negotiated and concluded over the telephone with papers sent back and forth for signatures. How it happens depends on the scope of the deal and the nature of the people that you are dealing with. Remember the personal dimension as you negotiate. Build a strong rapport with the key decision maker of the licensee company. Deals are rarely successful if there isn't one person within the company who champions your invention. If you have a good relationship, it will go far in helping you to iron out various terms such as how often you can inspect the company's facility, receiving samples, or the license's scope.

It is highly desirable to have a third party represent you in negotiations. Even when you are the one calling the shots and accepting or rejecting deals, it is more palatable when your agent says that the pot must be sweetened instead of you saying, "No, I want more." Using an agent also gives you the opportunity to sit back, listen, and think during the negotiation so you are fully prepared to make informed decisions.

Additionally, an agent allows you to play good guy–bad guy. For example, as an agent I might say that I disagree with a term in the agreement. In reality the inventor was against it, but by playing the role of the bad guy I take the heat off the inventor, who will have to work with the company in the future. It is important to select your roles ahead of time.

Manufacturers generally want to have their patent attorneys involved in drafting a license contract. In this case, it is best to have your patent attorney represent you.

WHEN TO MEET IN PERSON

If you are curious about the manufacturing facilities you targeted in your research, and you have the time and money to travel, it's worth it to visit whenever an executive is willing to meet with you. It's an opportunity to build a strong rapport, enhance communications, and gain more knowledge.

If you have six potential manufacturers scattered all over the country and no solid commitment of interest from any of them, then you may not want to bother with what will likely be a time-consuming and expensive endeavor. Wait until you have a solid expression of interest before attempting to meet in person.

On the other hand, a company may pay your travel and hotel expenses if there is a lot of interest. In fact, the offer to pay travel expenses is a good indication of the company's level of commitment. You know you've found a player when it doesn't hesitate to fly you in.

In my experience two visits are average for a licensing deal. The first visit is an opportunity for all the parties to meet, and the second is the time to sign the deal.

You are likely to be at a disadvantage when talking with the company's attorney. As a rule, you should have your attorney communicate with the company's attorney. Your attorney may find things that you would not. An ounce of prevention is worth a pound of cure, even when the prevention costs $200 per hour.

The License Contract

A license contract can be very simple. I have negotiated several license contracts that are one page in length, primarily one paragraph. The key elements in a license contract are:

- What is being licensed. This includes a patent if there is one, and any other proprietary information such as know-how or trade secrets.
- What the licensor is granting to the licensee. The inventor may be offering the right to manufacture, or to manufacture and market, or to market only, and so on.
- How much will be paid for the license: for example, a royalty of a certain percent of the manufacturer's gross income, or a dollar amount for every unit distributed.
- When the license fee will be paid: quarterly, annually, or otherwise.

Therefore, a very basic license contract could look something like this:

Sample Non-Exclusive License Contract

Inventor, John Paul Jones, hereby grants Acme Wadget Company the non-exclusive license to manufacture, have manufactured, and sell the Wing Wong Wadget, U.S. Patent No. 7,000,001, in consideration for Acme paying the inventor a royalty of 5 percent of all income received by Acme from the sale of said Wing Wong, and paid quarterly within 30 days from the last calendar quarter in which Acme received said income.

Agreed to this ________ day of __________________, 2015.

by

____

Inventor's Signature Date

also by:

____

Wilbur Ringer, President of Acme Wadget Company Date

Obviously a lot more could be stipulated in a license contract. There are a number of elements missing from this contract that many inventors want to address. There is no reference to the length and time of the license, the territory, performance standards, or escape clauses. (In the sample, the length of time of the license is understood to be forever because there is no limitation stated otherwise. The territory would be the world because there are no restrictions regarding boundaries.) However this simple contract contains the necessary terms for a valid license. It is not my intention to offer legal advice here, but to give you an understanding of the basic elements needed in a non-exclusive license contract. You should always consult with a patent attorney when considering entering into any license contract.

Know what points you're willing to give on.

Negotiating Contract Terms

When considering the various terms in a license contract, you need to give and take. Don't let the less important elements of the license contract get in the way of your overall deal. Don't be so adamant about inspection of your product, audits of

Simpler Can Be Better—Why would anyone want to negotiate such a simple and wide-open contract as the sample? The terms are so straightforward and simple that the key decision makers I was dealing with signed the agreement personally rather than referring it to their patent attorneys.

The anticipated royalties for the product under consideration were not very high. Had it necessitated a lot of wrangling with patent attorneys, the licensees would quickly have lost interest and determined that the up-front expenses for developing a license contract would have outweighed potential benefits. Potential litigation was out of the question because there was not enough at stake with any one company to ever warrant it. Interestingly enough, it was my patent attorney who suggested that I negotiate such a one-page license as a strategic ploy to entice several non-exclusive licensees to pay royalties. He reviewed my final draft. It worked and royalties flowed.

the licensee's records, copy text on the packaging, and so on, that the potential licensees throw their hands up and say, "Forget the deal." It's happened before.

If your invention will make a significant difference to the manufacturer's profits, then you are in a good position to call the shots. If your invention is one of many, then the reasonableness you show in your negotiations may go a long way in deciding whether a deal is eventually consummated.

When you are dealing with very large companies, the corporate patent counsel (attorney) will have special contract terms that the company likes to include in its license agreement. In this case, let the company initiate the first draft of a proposed license agreement. You save the cost of having your patent attorney draw up a license contract only to have it drastically changed by the company's corporate patent counsel.

Small to medium sized companies use patent attorneys only as the need arises. The company executive may instruct the patent attorney to draft the contract as quickly and inexpensively as possible. These patent attorneys may include boilerplate terms and conditions, some of which are unreasonable or inappropriate for your circumstances.

If instead you take the lead and present your terms and conditions to the company for presentation to its patent attorney, the attorney may accept your contract without adding other terms unless requested to do so by the company. Even though you may pay a little more initially for drawing up the contract yourself, you're in the driver's seat, and the negotiations may proceed more easily, less expensively, and in your favor.

When outlining the terms of a license contract, I propose each and every term ideally for my situation, then add a couple of terms that I could easily live without. I know ahead of time which of these terms I am going to give on during negotiations.

After you have presented all the terms of the license contract the way you would like them, the manufacturer will respond with suggested changes. As the negotiation proceeds, you'll come to understand which of the manufacturer's modifications are important to the company. Keep alert to those items that appear to be nonnegotiable.

For example, in one contract negotiation, my client was interested in maintaining control of the wording on the packaging. The manufacturer vehemently opposed this; it wanted complete control of the text and the design of the packaging. As is true for most manufacturers, its packaging and many of the terms on it needed to match the rest of the product line.

Another manufacturer was willing to compromise by allowing us to supply the text for the back of the package, as long as it met the company's approval. A different company denied our right to control the text on the packaging, but was willing to "consider" any input that we offered.

These are the kinds of back and forth negotiations that can drive both you and the manufacturer crazy. So be smart and don't let them get out of hand. Stand firm on your most crucial points, but let the minor ones go rather than running aground on them.

WORDS OF WISDOM

As an Army intelligence officer once said, "It matters not what the facts are, only what the facts are perceived to be." And, according to Stephen Covey, "Seek first to understand, before seeking to be understood." Bottom line: make sure you know the perspective of the other side in negotiations, and to paraphrase True Value Hardware founder John Cotter, "Say no more than necessary to keep them talking."

Contract Terms

The following pages describe many of the most common contract terms. The purpose is not to offer legal advice, but to help make you aware of the various types of business decisions you must consider and the effect they have on a license contract. This is no substitute for having a qualified patent attorney assist with the legal aspects of your license contract.

A contract begins with a title, such as Contract, Agreement, or License Agreement. The first paragraph identifies the parties involved in the agreement. In the section that begins with Witnesseth, the paragraphs that start with Whereas explain the reason(s) for entering into the agreement, i.e., who approached whom, what the objectives of the parties are, and what they expect from each other. These are called the recitals. After the words "Now, therefore . . ." and "agreed as follows: . . ." are the actual meat and potatoes of what is being agreed upon.

Roll with the punches; retreat is not the same as surrender.

One of the primary purposes of having a written contract is so someone completely unfamiliar with the two parties can understand what the parties agreed upon. After all, what if one of the parties dies or goes out of business? The agreement should spell out what is being agreed to in easily understandable terms.

License Agreement

AGREEMENT (hereinafter "Agreement") made as of December 13, ________ by and between __________________________ having an office at_____________________ incorporated under the laws of ___________ (also sometimes herein referred to as "Licensee") and RONALD L. DOCIE, residing at ___________, Ohio ("Licensor").

WITNESSETH:

WHEREAS, Licensor is the owner of the invention and patent described in U.S. Patent number ______________, entitled _____________ (the "Patent"), and

WHEREAS, Licensor owns the exclusive and transferable rights to said Patent and has the right to license same, and

WHEREAS, Licensee desires to obtain a license for said Patent from Licensor, and Licensor is willing to grant said license to Licensee upon the following terms and conditions, it is

NOW, THEREFORE, in consideration of the mutual promises herein contained, agreed as follows:

1. The Invention: The term Invention as herein used shall include the U.S. Patent number _________, the teaching contained in said patent, all improvements, modifications, and changes in said Invention, know-how, and trade secrets in any way relating to the blind spot mirror wedge now in existence, in development or hereafter developed at any time during this License.
2. Representations and Warranties: Licensor hereby represents and warrants that he possesses the sole, exclusive, and unrestricted ownership in said Patent and has the sole, exclusive, and unrestricted right to license same. Licensor hereby indemnifies and agrees to hold Licensee free and harmless from any claims or liability with respect to Licensor's rights to grant this License.

   Licensee agrees to use best efforts to exploit the Invention. For lack of diligence in any manner herein specified. Licensor shall have the option to terminate this Agreement according to Article 21.
3. Grant: Licensor hereby grants to Licensee a non-exclusive license to make, have made, use, and sell the Invention for the term set forth. This Agreement is not transferable nor may any Licensee assign or sublicense same without the express written consent of Licensor, except as to the transfer of that portion of Licensee's business as to which this Agreement applies.
4. Territory: The territory of the license of this Invention is the United States and its possessions, which territory may be expanded with the approval of licensor.
5. Royalties: Licensee agrees to pay to Licensor a royalty of five percent (5%) of the net selling price on revenues received from sales of the Invention. Net selling price, for purposes of this Agreement, shall mean the gross sales price less actual returns and normal trade discounts, if any.
6. Time of Payments: The royalties contemplated in Article 5 shall be paid within thirty (30) days after the end of each calendar quarter, for royalties payable during said calendar quarter. The first calendar quarter shall be the fourth quarter of the year _____. Royalties due and payable September ____ shall be paid as part of the fourth quarter of the year.

7. Reports: Licensee shall supply Licensor, at Licensee's expense, a written report of the number of sales of the Invention, by model, within thirty (30) days after the end of each calendar quarter, for sales made during said calendar quarter.
8. Initial Fee: An initial fee of ______, payable to Licensor, shall be returned with the signed copies of this Agreement if returned on or before January 15, ____. If returned after January 15, ____, said initial fee shall be ______.
9. Duration: Royalties shall commence as of September 1, ____ and, unless terminated under Article 21 hereof, this Agreement shall remain in full force and effect for the life of the Patent and last to expire of any patent applications included herein.

   In the event Licensee discontinues sales of the Invention for a period exceeding four (4) calendar quarters during the term of this Agreement, Licensor may terminate this Agreement according to Article 21. In this case, past due royalties, if any, shall become due and payable according to Article 19.

   In any case, for this Agreement to be valid, said Agreement must be signed by Licensee in duplicate and returned to Licensor by February 28, ______.
10. Overdue Payments: Overdue payments to Licensor from Licensee shall accrue interest at the Prime Rate announced from time to time by Chase Manhattan Bank. Interest payments are due along with the overdue payments.
11. Packaging: Licensee may use any style, shape, size, and design of packaging for the Invention consistent with good taste and Licensee's normal packaging. Licensor reserves the right to include descriptive copy and drawings, describing the proper use and advantages of the Invention on the packaging, and limited on the face of said packaging so as not to substantially interfere with Licensee's overall theme.
12. Specifications: Licensor shall, within ninety (90) days of the execution of this Agreement, supply drawings, specifications, teachings, and know-how to Licensee who shall manufacture the product in reasonable accord therewith. Licensor shall have the right at his expense to inspect Licensee's facilities and product for quality control once each calendar quarter.
13. Marking: Licensee agrees to display on the product, or on the package in which the product is sold, the following: "Made under license of U.S. Patent No. _________." It is understood by the parties that should any improvement inventions be included in this Agreement, such patent numbers would be added to this notice.
14. Enforcement: Licensor shall use due diligence to enforce patent rights under the Invention against infringers. Licensee shall provide Licensor with any and all information that may assist Licensor in this regard. All expenses relating to enforcement shall be borne by Licensor, and Licensor shall be entitled to the entire recovery, if any. In the event Licensor fails or refuses to diligently prosecute any such infringer, and when such failure adversely effects Licensee's revenues, Licensee may terminate this Agreement according to Article 21.
15. Trade Secrets: If Licensor provides Licensee with information, and/or new developments patentable or/not, and stipulates "Trade Secret" or "Confidential," Licensee shall use discretion to not disclose said information except on a need-to-know basis and until such time that Licensor gives Licensee written permission to release said information. This paragraph does not apply to information that is already in the pubic domain or information already known by and documented by Licensee.

16. Interference: Licensee shall not use any information provided by Licensor to provoke interference with Patent and any patent applications which Licensor has or may file, and said information shall not be used to amend any claims, in any pending patent applications, to expand claims to read on, or cover, or to dominate, any inventions or submissions of the Licensor whether or not patentable.
17. Non-Compete: Licensor shall not directly compete against Licensee or Licensee's accounts for products under the Invention. However, this shall not in any way limit Licensor's right to grant other non-exclusive licenses to others.
18. Samples: Licensee shall supply Licensor with twelve (12) samples of each model of the Invention, at Licensee's expense, each calendar quarter, upon Licensor's request. In addition to free samples, Licensor may purchase products of the Invention from Licensee, over and above those products made for Licensee's own use or sale, at five percent (5%) under Licensee's lowest price sheet price. No royalty is to be paid on any samples supplied or sales to Licensor.
19. Royalties on Previous Sales: Licensee is relieved of paying royalties on past sales of the Invention made prior to September 5, _____, if any, unless, upon termination of this License grant, royalties paid to Licensor during the term of this license do not exceed the royalty that would have been received by Licensor from past sales of the Invention according to Article 5. In this event, the difference between royalties paid and past royalties due would be due and payable to Licensor.
20. Indemnification: Licensee shall indemnify and hold Licensor harmless from any damages, claims, costs, expenses, or suits arising out of or incurred in connection with Licensee's actions with respect to third parties including, but not limited to, customers of Licensees and customers of Licensee's customers.
21. Termination: In the event that either party breaches any term of this Agreement, the other party shall have the right to terminate the Agreement thirty (30) days after the non-breaching party notifies the breaching party in writing of said breach, providing the breaching party fails to cure said breach within thirty (30) days.

    In the event this License is terminated, Licensee shall have the right to liquidate its inventory of the licensed product, then on hand, on condition that Licensee continues paying royalties on all sales thereof, as herein before provided. Licensee agrees to not sell or otherwise exploit the Invention thereafter without express written permission of Licensor.
22. Assignment: This Agreement shall bind and inure to the benefit of each of the parties hereto, their successors, heirs, legal representatives, executors, and assigns.
23. Waiver: The waiver of a breach hereunder may be effected only by a writing signed by the waiving party and shall not constitute a waiver of any other breach.
24. Modification: The parties agree that the present contract may not be modified or amended except by a writing signed by all parties, which said writing shall refer specifically to the present Agreement and state that it is amending or modifying same.
25. Applicable Law: This Agreement shall be construed and enforced in accordance with the laws of the State of Ohio, except as to any provisions thereof that are governed or governed primarily by the laws of the United States of America, in which case the latter laws shall govern.
26. Arbitration: Any controversy or claim arising out of or relating to this Agreement, or the breach thereof, shall be settled by arbitration in accordance with the rules of the American Arbitration Association, and judgment upon the award rendered by the Arbitrator(s) may be entered in any court having jurisdiction.

27. Reform: The parties agree that if any part, term, or provision of this Agreement shall be found illegal or in conflict with any law, the validity of the remaining provisions shall not be affected thereby.
28. Notices: Any notice that is required by the terms of this Agreement to be given by either party to the other shall be deemed to have been given if and when properly deposited for sending by registered mail, prepaid, return receipt requested, and properly addressed to the other party at the address stated above, or at such other address as either party may supply to the other in writing duly acknowledged by the addressee.
29. Records: Licensee shall keep accurate records of all operations affecting payments hereunder, and sales and market information relating to the Invention and shall permit Licensor, or his duly authorized agent, to inspect all such records, during regular business hours throughout the term of this Agreement and for a period of two (2) years thereafter. This right shall be exercised no more than twice during any calendar year. Licensor acknowledges that Licensee's books and records contain proprietary information of Licensee. Licensor therefore agrees to keep any such information confidential and use same only for the aforesaid purpose.
30. Entire Agreement: This Agreement contains the entire understanding between the parties and supersedes any prior understanding or agreements between and among them respecting the subject matter. There are no representations, arrangements, understandings, or agreements, oral or written, related to the subject matter of this Agreement except those fully expressed herein.
31. Captions: The captions used in connection with the paragraphs and subparagraphs of this Agreement are inserted only for the purpose of reference. Such captions shall not be deemed to govern, limit, modify, or in any other manner affect the scope, meaning, or intent of the provisions of this Agreement or any part thereof, nor shall such captions otherwise be given any legal effect.

IN WITNESS WHEREOF, the parties have set their hand and seal as of the day and year first above written.

In the Presence of:

______________________	______________________	
Witness	LICENSOR	
______________________	______________________	__________
Witness	by	Date
______________________	______________________	
Witness	LICENSEE	
______________________	______________________	__________
Witness	by	Date

Contracts used to be so complicated that only an attorney could understand them. However, following a contract reform movement even large companies are modifying contracts so anyone can understand them. The sample contract falls somewhere between easy to read and attorneys only. It is replicated from an actual agreement that I negotiated, with legal assistance from my patent attorney.

THE INVENTION

Most license contracts will define the term *invention*. For example, are you offering the use of a patent only, or are there also know-how, trade secrets, and modifications of your technology? One of my clients invented an electrical device and obtained a U.S. patent. This patent represented just one of several new products the same inventor developed in that field. He believed that he was on the cutting edge of developing new and more progressive electrical technologies along the same line that would surpass the technology described in his existing patent.

This inventor might be well served by limiting the initial license contract to include only the existing patented technology. This way, when he develops new and improved technologies, they would be the basis for an additional license contract. His other technologies might command even higher minimum performance standards. This would also leave the door open for the inventor to license his new technologies to a different company if the existing manufacturer did not capture market opportunities as anticipated.

If your invention incorporates a basic concept that you believe is not subject to much more development, you may want to expand the definition of invention to include know-how, trade secrets, improvements, modifications, and improvement patents. If the manufacturer slightly modifies your invention or adds improvements to it, these variations may still fall within the scope of your license contract and you could still collect royalties on them.

REPRESENTATIONS AND WARRANTIES

These assure the licensee that you are indeed the rightful owner of the proprietary rights that you wish to grant a license for. Licensees commonly want you to indemnify them and hold them harmless in the event that something beyond their control occurs. For example, an inventor's former business partner might claim to be the joint inventor of the technology and decide to sue both the company and the inventor. Obviously, the company isn't at fault in this matter, and there's not much it can do to help resolve the situation. Therefore, it would expect the inventor to handle any expenses involved in such actions. This puts the burden on you to be sure all your ducks are in a row before entering into a license agreement.

In the sample contract I added a paragraph to this clause regarding the diligent manner in which the licensor needs to safeguard this product. Failure to do so would be grounds for termination. Although this clause could have been included separately, it is not uncommon for clauses to be combined during negotiation. Last-minute changes may also be hand written on the contract; both parties must initial and date the page where a change has been made to express their acceptance of the change.

GRANT

The term *invention* refers to the intellectual property you are offering to the licensee (what it will get). The *grant* addresses what it is you are allowing the licensee to do. Here you specify whether it is an exclusive or non-exclusive license grant, and whether the licensee has permission to manufacture your invention, use it for its own benefit, and/or sell it to others.

The grant also specifies whether the license is transferable from the licensee to another company. In other words, will you allow this licensee to re-license another company to perform the same functions under this license? Your licensee could essentially sell or otherwise offer this license contract to another company. That company would normally have to perform under the same terms and conditions of this contract, but would it be as effective? I advise that you avoid this scenario if possible.

A good compromise would be to offer the following term: "This license is not transferable by licensee except with the transfer of that portion of licensee(s) business to which this license agreement applies." This means that if your licensee chooses to sell that aspect of the business that relates to your products or technologies, it would be allowed to transfer this license in conjunction with such a business sale. You are not limiting your licensee if it chooses to change its business interest. If the company is insistent on having such a condition in the contract, you may want to include a clause such as this one.

TERRITORY

The territory clause is a significant one. Will you offer your licensee the territory of the United States, North America, the world, or some portion thereof? If your manufacturer has a strong market position in the United States and a weak position in Europe, why offer them the world? Generally you would not want to offer any greater territory than is necessary.

Europe can be a very sizable market. If several European companies control the vast majority of the European market, you would want to limit the territory of the U.S. manufacturer to the United States or North America, at least initially. You can always offer European rights later if the U.S. manufacturer performs adequately in the United States, which provides a performance incentive.

However, your manufacturer may have a strong market position in the United States and an interest in aggressively pursuing the European market. It may only deal with you by having rights to the United States and Europe. This scenario puts you in a strategic bind. I would probably select the U.S. manufacturer and offer them both U.S. and European rights. However, since your U.S. manufacturer has not yet demonstrated a substantial track record in Europe, it would be smart to carefully define performance standards and a time limitation specifically relating to European sales.

PERFORMANCE

This particular clause does not appear in the sample contract because it is an example of a non-exclusive agreement. Performance clauses are generally only included in exclusive agreements. This section addresses what will happen if the manufacturer fails to meet its performance standards.

If it fails to meet performance standards you could completely rescind its marketing rights or rescind its exclusive position, thereby giving you the opportunity to license to other companies that may be in a stronger position to capture that market. If the company rolled out a million-dollar budget to introduce your invention and failed to meet the performance standards but came close, it certainly will not want to give up its rights altogether. If, however, it only lost its exclusive position, it could keep marketing your invention. It may not be unreasonable for you to give the licensee a break if it puts forth real effort.

Another potential standard for performance is to have the company commit certain monies or resources to your project for a specified period of time. Have the contract specify minimum expenditures and/or minimum shipment performance. Minimum shipment means it must actually sell products, not shelf the project and pay you the minimum royalty just to keep others out of the market.

ROYALTIES

This is where you would define your royalty rate, including any specific deductions such as returned products, shipping charges, rebates, sales commissions, or other such expenses that are not directly associated with the manufacture of your product. Deducting these items changes the basis for payment from gross sales to net sales.

TIME OF PAYMENTS

Payments are normally quarterly, sometimes monthly, or less commonly annually or otherwise. You will be hard pressed to receive royalties any sooner than thirty days after the end of the calendar period in which royalties are due. Some licensees like to extend this to forty-five or sixty days. This term often depends on the company's accounting system and the strength of what you have to offer. In the long run, timely payments can make a big difference to your cash flow.

REPORTS

You might find a reports clause toward the end of a contract. I put it after time of payment simply because I expected reports to coincide with payments. The order of these clauses usually doesn't make a difference; however, it is normal to start with the more substantive terms.

It is my experience that most companies include a report of sales; they have to generate it for themselves anyway to establish your royalties for that period. If there are different models or sizes of your invention, this is where you would ask the company to give you a breakdown so you can clearly track your sales performance.

INITIAL FEE

We inventors generally want to receive as much money up front as possible. If nothing else, we would like to be reimbursed for the expenses we paid out for prototyping, patenting, and development. It is nice to have hard expenses covered, and even better to be reimbursed for our time on the project. More times than not, this is not an option. Companies are reluctant to pay up-front fees for licenses. After all, if they need to pay a lot of money up front, they could have purchased the proprietary rights outright, rather than licensing them.

If a company won't agree to a large initial fee, suggest an initial fee payable over a given period of time. This could be quarterly or at certain milestones, such as when final working models are developed, after a sale to a major account, following introduction at a major national or international trade show, or at the first shipment. This way the company does not have to pay out large sums of money until the project meets expectations and shows future promise.

The more common scenario is for the licensee to agree to a modest initial fee, enough to cover your out-of-pocket expenses for development thus far. This initial fee may be deducted from future royalties. The company may not start royalty payments until the entire initial fee amount is deducted. It would be preferable to negotiate a proportional deduction whereby a certain percent of your initial fee payment would be deducted until the full initial fee amount was accounted for. If the company resists such a complicated term in an agreement, it is probably due to its standard accounting practices and the inconvenience of making modifications to it. If you are negotiating a large deal that is valuable for the company, you have more leverage to insist it meet your contractual desires. However, if in the greater scheme of things this is a small deal, the company may be less willing to modify something like its standard accounting procedures. Remember to give and take; don't let an issue like this kill the deal.

RAMP-UP TIME AND PENALTIES

Although not shown in the sample contract, this would be a good place to include such a clause if it were appropriate for your situation. The ramp-up time is the length of time it takes to get your product or technology market ready. The anticipated ramp-up time may affect your initial fee or the potential need to impose penalties. Generally speaking, the longer the anticipated ramp-up time, the greater an initial fee is warranted. This fee may come in one lump sum or in payments over a period of time. It is possible that by the time your invention reaches the end of the ramp-up period, your licensee may decide to not proceed further with the project. This happens all too frequently. The result is that you lose a tremendous amount of opportunity, in both time and money.

A manufacturer might eventually pass up your project for many reasons. It might sell off part of its company to a firm that is not interested in pursuing your product line, there could be a management change with the associated repositioning, or any number of market factors and other externalities beyond your control could occur.

An incentive for licensees to proceed diligently with your project, without requiring them to make payments along the way, is to tie nonperformance to a penalty. For example, you could insist that the company pay you a certain amount of money if it does not reach a certain stage of development within a certain period of time. Another penalty may be that the company would lose its exclusive license grant and convert to a non-exclusive status if it does not reach a certain level of development within your target time limit. A third type of penalty may be tied into minimum investment increments agreed to by the company.

DURATION

How you define duration depends on how you defined your invention. If you are throwing in the kitchen sink, licensing your entire intellectual knowledge including know-how, trade secrets, modifications, and so on, you may want to expand the duration to the greatest extent possible and to include any additional patent applications made relating to your invention. This is what is meant by "patent applications included herein."

Let's say that you have thirteen years of life left on your existing patent and in ten years you make an additional proprietary development and improvement to your invention. This may be the subject of an additional patent application. If you receive a patent on this improvement, and if the company wants to continue to produce your invention based on this, you could essentially extend your royalty revenues for another ten to twenty years, depending on the overlap between the coverage of the two patents.

On the other hand, if you believe there is a real possibility that you may develop additional technologies that would be appropriate for other licensees than the one you are currently dealing with, you would be limiting yourself to tie these additional developments into your existing licensee. Such a clause would lock you into your existing licensee for a very long period of time. This can be good or bad, depending on the potential for your own continued development and on the position that your licensee maintains in the industry. A lot can change in ten to twenty years.

Such a simple term as duration can potentially have a huge effect on what you or your heirs will eventually receive.

OVERDUE PAYMENTS

It never hurts to provide some incentive to the licensor to make timely royalty payments. When a well-run corporation is in a cash crunch, its accounting department will sometimes first pay those bills that have the highest interest and penalties associated with them. Perish the thought that this would happen in your situation; however, if it did, such a clause may assure your payments, or at least offer reimbursement when they are late.

Approach any deal with vulnerability—but don't show it.

PACKAGING

As I mentioned previously, trying to stipulate packaging criteria to a manufacturer can be a deal killer. However, if there is certain copy or an explanation about your product that you feel is particularly important, you may want to be very specific about what will be included on your packaging. For example, you may want to insist that the licensee include at least a three-square-inch diagram that shows how your invention works.

SPECIFICATIONS

List any limitations that you want in place regarding the exact specifications of your technology, such as specific sizes and dimensions of your product, and whose responsibility it would be to set these specifications. Would it be your obligation to supply the company with working drawings and exact specifications that would not be altered, or would you offer the licensee the flexibility to design its own specifications and make modifications to your designs?

You may require that your licensee develop specifications within a certain period of time and that such specifications must meet with your approval. This gets into a sticky area because many inventors want tight control over the specifications of their inventions while licensees like to have the flexibility to make changes.

I believe it is important to be flexible here and to offer some compromise. I have seen arguments over this term alone end up squashing deals. It helps when you have the opportunity to talk directly with your licensees' engineers and designers to make sure you are all on the same page. This alone can greatly reduce disputes.

MARKING

There are important reasons for making sure the patent number appears on your product and its package. If the patent number is missing, it invites competition from foreign manufacturers that believe they can legally plagiarize your invention. It is your duty as the inventor to put the rest of the world on notice that your product is indeed covered by a patent(s). If someone infringes on your patent, you may not be able to claim damages unless you have the proper marking on your product.

Too often we forget that genius . . . depends upon the data within its reach, that Archimedes could not have devised Edison's inventions.
—ERNEST DIMNET

You would think that such a clause would be obvious, if not unnecessary. However I have found it extremely helpful to include such a clause. In today's age of products being manufactured overseas, it is easy for a licensee to become lazy with respect to specifications on packaging. Even though you would think it is in the licensee's best interest to include patent numbers, they don't always do it.

ENFORCEMENT/INFRINGEMENT

In the sample, this clause states that the entire burden of responsibility regarding enforcement lies with you, the licensor. There are many variations to this term in a contract. For example, you may do just the opposite, making it solely the licensee's responsibility to pursue any patent infringement, in which case the licensee may be entitled to the entire recovery, if any. A variation of this would be to make it the licensee's responsibility to pursue infringement, but to specify that you, the licensor, would be entitled to a portion of said recovery, perhaps in an amount equivalent to your royalty rate. Therefore, if the licensee were to recover $100,000 in an infringement settlement, and your royalty rate was 5 percent, you would receive $5,000 of the recovery.

I prefer another variation. It is a compromise between the previous two scenarios and is particularly appropriate with non-exclusive license grants. You, the licensor, would be responsible for initiating any enforcement action against infringers, and you would be entitled to the entire recovery, if any. Additionally, if you chose not to initiate such an infringement action, the licensee would be entitled to initiate such an enforcement action on his or her own. This would be advantageous for a licensee if the infringement was not a critical issue for you, at least not enough to warrant investing several thousands of dollars. It may be important for your licensee to initiate such an enforcement action so that it can set a precedence on this issue. Such a precedence may effect the licensee's other products. Its stakes may be much higher than yours.

You could have as part of your terms that any proceeds from the recovery would be split proportionately based on the contributions each party makes. After investing $20,000 to initiate patent enforcement proceedings, you may find yourself at the end of your financial rope, unable to go any further. Your licensee may see enough merit in the case to proceed and contribute another $200,000 to the patent enforcement case. You would recover 10 percent of the settlement, and your licensee would recover 90 percent.

TRADE SECRETS

Identify how trade secrets will be dealt with and to what degree your licensee is bound to secrecy if you reveal trade secrets or additional nonpatentable information. You may be required to specify trade secrets in writing and stamp them as confidential to put the licensee on notice and to put this clause into effect.

INTERFERENCE

Some major corporations want to reserve the right to interfere with your patent. They will attempt to invalidate your patent by trying to prove that you are not the first to invent such an invention. Basically, this clause states that the company will not mess with your patent position, especially by using information it may have gained from you. You might think this is a pretty crafty thing for a company to do, and the question would be why wouldn't it agree to such a clause. If it is a very large company with an active research and development department, it may need to steer a clause like this in their favor to avoid compromising an existing position with one of its subsidiaries or an in-house inventor. After all, you may not be the original inventor, and there may be a company employee who did indeed conceive the technology prior to you but failed to make management aware of it. By avoiding this clause as it's written in the sample contract, the company safeguards itself from such a scenario and protects itself from unscrupulous inventors who file superfluous patent applications in an attempt to coerce the company into unjust payments. This can all become very complicated, which is why I rely on the advice of patent attorneys for clauses like this. Again, this is one you could be flexible about.

NON-COMPETE

This term is normally straightforward, but there are exceptions. You may have created a small niche market for your invention, one that is profitable and ongoing. You may want to retain the rights to continue to sell your invention in your niche market, at least until your licensee is up and running on a major national scale. This can be important if you are not able to establish large initial fees and it will be some time before you anticipate royalty payments. How much you give and take on this term depends on the commitment your licensee is willing to make and how much you stand to lose if you have to pull out of your market.

You could also set time limits and due dates by which your licensee would take over the entire market, at which time you would cease competing. This is subject to negotiation and compromise. Remember the effect that your market presence may have, no matter how small, on diluting your licensee's market position. If your licensee is preparing to create name recognition for your product, your presence in the market may be confusing and could adversely affect the licensee's introduction of your product, thus reducing their enthusiasm and your royalty income.

SAMPLES

I always like to include free samples for inventors as a term in a license contract. This is one of the best ways to maintain some degree of quality control over your invention. Manufacturers typically send out samples for sales prospecting. Even though the manufacturer assures you that it will send you samples from time to time, it's best to include in your license contract exactly how many samples you will receive and when. With a low-cost invention, you may want to request a dozen free samples to be sent with your royalty payments each quarter. With larger or more expensive inventions, one sample per year may be appropriate.

Sometimes manufacturers prefer to make it your responsibility to request samples. Their computers are set up to remember royalty payments but not to send out samples. I believe this is a reasonable term to accept. In some cases it will be specified that your request for samples must be made in writing. Some manufacturers insist that the inventor buy the samples at cost or at the highest volume discount price. Samples are normally royalty free, meaning the manufacturer will not pay you royalties for samples of products that you receive.

ROYALTIES ON PREVIOUS SALES

This is not a common clause because normally you are not negotiating with a licensee that is already selling your product; however, it does happen. It is my understanding that companies would not owe you royalties on products upon which they are infringing until you put them on notice that they are infringing. Patents are generally considered defensive in nature. It is your burden to prove that a company is infringing on your patent and then your responsibility to notify the company as such. The sample contract refers to a situation in which the licensee certainly had knowledge of the fact that there were patent rights on my invention prior to my approaching the licensee, but proving it was difficult. Since they knew that I knew that they knew, the company was gracious enough to offer a royalty on previous sales as outlined in the sample contract. I was able to accomplish this feat in about half of the instances where this arose.

INDEMNIFICATION

Indemnification or hold harmless clauses are often resisted by whomever they are applied to. They put the entire burden of responsibility on the other person to pay for all legal actions and damages made to the person who is held harmless. Obviously licensees like to avoid this when possible. But the licensee is representing your products, possibly to your specification. If the licensee modifies your specifications or does not maintain proper quality control, you do not want to be held responsible for damages that result. Therefore, it is reasonable that you be held harmless from the negligent actions of your licensee.

TERMINATION

Termination clauses come in many forms, as you might imagine. The clause as shown in the sample contract has its obvious advantages. If one party makes a mistake or reneges on any terms of your agreement, it has the opportunity to correct its mistakes. This puts the burden on each party to notify the other regarding any such breach. First you have to decide whether a breach is serious enough to warrant termination of your contract. You may want to overlook certain minor breaches and give the company more time to correct breaches if there are extenuating circumstances. You can stiffen this term of the license contract if you prefer; such terms are largely negotiable between the parties' attorneys.

ASSIGNMENT

This clause is designed to describe who may end up with the rights garnered under this contract in the event that something happens to one of the parties, such as death. Spelling it out like the sample contract does leave it somewhat open for both parties to decide on their own who might inherit the assets and responsibilities.

WAIVER

One party can waive any part of the terms and conditions of your contract. To do so, it must be in writing. If the company didn't quite make its minimum performance standards, but it showed good effort and you have no other alternative for licensing, you may choose to allow the company to continue with an exclusive arrangement. The way this sample contract is written, waiving such a right does not mean that you are waiving any of the other terms and conditions of the contract. And it doesn't mean you'll automatically waive this right again.

MODIFICATION

This clause spells out the terms under which the parties may modify the contract at some point in the future. In the sample contract, any such modification or change must be in writing, it must refer specifically to the agreement, and it must be referred to as an amendment. If a future modification is necessary you should make specific reference to the name and date of any agreement you are modifying.

APPLICABLE LAW

This is one of those clauses that you could live without because if you and the other party are in different states, regardless of what you put in this contract, an attorney may argue some reason to establish a lawsuit in whatever state is more favorable to the attorney's client. However, if you can get a clause like this stating that it be governed in your state, it could save you money if you ever need to litigate.

ARBITRATION

There are pros and cons about whether to put an arbitration clause in a contract. Certainly an attorney can advise you better about this. It is thought that arbitration can save you money if there is a dispute over the contract because arbitrators may resolve a conflict and help you avoid a lawsuit. A clause like this in your agreement does not guarantee that you would avoid a lawsuit, and arbitration is not always cheaper.

REFORM

Reform is explained in the clause of the sample contract. It means that if part of your agreement is thrown out of court because it is somehow improper or not written within the bounds of the law, the remaining terms and conditions of the contract shall still be in full force and effect.

NOTICES

The terms for this clause are quite simple. This clause outlines the proper procedure for notifying the other party for things such as potential breach of contract, change of address, and royalties were not received.

RECORDS

Inventors sometimes worry about whether a manufacturer is paying royalties for the appropriate number of sales. The problem is that you cannot waltz into the manufacturer's accounting department and ask to see its books. Manufacturers are very protective about their accounting records. They do not want any sales information leaked out to their competition. Therefore, manufacturers will generally insist that any inspections you make of their accounting records be done by a Certified Public Account (CPA). And it ain't cheap. They will also normally ask that you bear the cost of such an inspection. This goes a long way toward limiting inspections of their accounting records. A CPA would be charged with maintaining secrecy about information that is not relevant to you. You would only receive information related to your product(s).

ENTIRE AGREEMENT

This clause specifically states that your agreement shall supercede any prior understanding (made in writing, over the phone, or in a face-to-face meeting) so this is the only formal agreement between the parties unless modified according to the terms of the contract's modification clause. Be sure that you are comfortable with all the terms in your contract, because any prior verbal understanding or written agreement will likely have no legal standing.

CAPTIONS

The captions are subheadings of the various terms and clauses in the contract and have no legal meaning in and of themselves. I understand that there have been lawsuits won and lost based on how the subheadings were interpreted irrespective of the content of the term following it. This clause avoids such confusion.

SIGNATURES

It has been my experience that most contracts are sufficient with each party's signature witnessed by one or two individuals. Sometimes attorneys advise using a notary. It depends on the governing laws of your state or country and what is at stake between the parties. Many of my license deals were done without witnessed signatures.

Let's face it. If someone really wants to back out of a deal with you, or give you a rough road to hoe in the process of trying to renege, there's not much you can do about it except lose lots of sleep and make a bunch of attorneys happy. Keep in perspective your potential profits versus the costs to protect it.

TRICKS OF THE TRADE

Sales Don't Lie—Although you may have provisions in the license contract for auditing the licensee's accounting books, I have found the best way to monitor licensees' sales is to become familiar with the distributors and retailers that are handling your products. Ask them about how many of your products they sell and extrapolate the data as necessary. Match this against your quarterly royalty statements.

An advantage of this is that you stay abreast of any changes in the marketplace. You will know when you need to develop improvements for your technology and to guard against infringers that your licensee may have missed. If a company wants to be dishonest and create a second set of books, your study of its books won't necessarily be the best way to find the deceit. Your searching and doubt will create friction between you and the licensee, not to mention your significant cash outlay for a CPA.

Monitoring licensees can keep you busy, but your diligence will pay off in spades.

Other Terms

There are numerous other terms that are not in the sample contract. Here are a few to consider.

FIELD OF USE

The field of use clause limits the scope of what a licensee can do with your product, addressing the issues of divided territory, delegated function, separate markets, or other situations whereby you restrict the licensee's authority to make, use, or sell your invention. See Chapter 5 for a more detailed discussion about when and why you might do this.

INSPECTION OF THE MANUFACTURING PROCESS

For highly technical or scientific inventions it might be important to be able to inspect the plant in which your inventions will be manufactured to make sure the intended materials and ingredients are used properly. Generally manufacturers don't want this term in a license contract, and they rarely allow frequent inspections. You may want to inspect quarterly, and the company will suggest a biannual inspection. A compromise, such as an annual inspection, may be reasonable.

QUALITY CONTROL

Terms outlined in a quality control (QC) clause may differ from Inspection of the Manufacturing Process and Specification. You may include provisions for monitoring the QC of your invention without necessarily inspecting the manufacturing process. You may have developed a set of specifications for your product, but the manufacturer will ultimately create its own specifications.

A QC clause could involve a number of variations. For example, you could have samples of the product sent to you so you could test them yourself, or you could require the manufacturer to supply you with a written report of its QC records, which would include date, time, and frequency of testing, along with the test results. With more sophisticated inventions there may be a provision for outside consultants to periodically monitor the QC of your product, in which case you would need to specify who would be responsible for covering the cost for such a consultant. You may also need to make provisions to cover a situation whereby the company repeatedly fails to meet QC standards.

MAINTENANCE FEES

If the license is exclusive, I recommend that the licensee be responsible for maintaining any patent maintenance fees payable to the USPTO. This would include fees for any of the patents that the licensee is utilizing under the license.

When you are dealing with several licensees on a non-exclusive basis, you are in a less advantageous position to request that any one of them pays for patent maintenance fees. An exception would be if any of your licensees wanted to negotiate on some other term of your license contract. If all but one of your licensees are willing to pay an initial fee, you could request that the holdout company pay for the maintenance fees on your patent in lieu of initial fees. That company can extend its cash flow, and if your product does not meet its expectations, it has a way of getting out of the entire deal with less financial exposure.

SPECIAL TERMS

There are many other possible aspects to a license contract that might be important in your situation. For example, if you anticipate doing additional research and development with your licensee's engineers, who will hold the patent rights to the new developments? Consider all the specifics of your situation that should be in your license contract, or develop additional contracts as appropriate.

Who Will Determine the Terms and Conditions?

Many people think that patent attorneys determine the terms and conditions of a license contract. However, a patent attorney will not normally know all of the strategic factors that need to be considered about your invention with respect to territory, scope of license, performance, duration, royalties and fees, and all the other critical elements that you have carefully weighed as a result of your extensive research.

Nor will a patent attorney normally have the expertise to delve into intense market strategy and product positioning. Therefore it is up to you, or your marketing and licensing representative, to define the parameters your patent attorney should work within.

Even if neither you nor your patent attorney have this level of information about your market, you can still negotiate a license, and a good one for that matter. A savvy patent attorney with experience in negotiating will simply determine the potential licensee's desires, ascertain which terms are negotiable and which are not, and then try to strike a deal as close as possible to what you want.

In this chapter we've extensively reviewed the technical elements that go into license contracts. It is important to be thoroughly familiar with these points, but the human element is also very important. Whom do you deal with, and how do you deal with them? Negotiation occurs between people, and understanding the other party's perspective is as relevant as knowing your own. As Stephen R. Covey suggests, "seek first to understand before seeking to be understood."

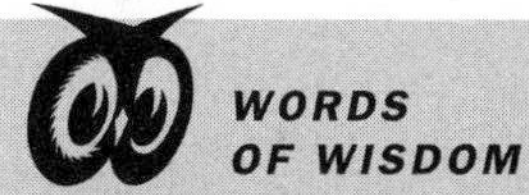

It is important to understand that sometimes you do not have much choice about the terms and conditions of a license contract. Most of the terms and conditions may be nonnegotiable. You may have only one company interested in licensing your invention. In this case, set aside your market research findings and take what you can get!

My successful negotiations for the Docie wedge mirror have been between myself and the presidents of companies—not their attorneys and not their vice presidents, but the actual decision makers. Sometimes people start negotiations by having their attorney talk to the other party's attorney. This is not necessarily the best way to go. In fact, I think doing so would have met with certain failure for the Docie wedge mirror. That negotiation needed a savvy licensing agent who thoroughly understood the intricacies of the industry and had a willingness to look beyond a strict legal perspective.

THINGS TO REMEMBER AND CONSIDER

- Practice visualizing the best possible scenario for a licensing deal.
- Identify the elements of a licensing contract that are most important to you; then identify those areas about which you will be flexible.
- Recruit appropriate professionals like patent attorneys or licensing agents to help you with the negotiation process.
- Contract points to consider: Exclusive or non-exclusive? Minimum levels of performance?
- Strategies to consider: Who will draft the first version of the contract?
- Who will negotiate the contract?

Ginning cotton by Eli Whitney (U.S. Patent No. 72X), New Haven, Connecticut,1794.

CASE STUDY
An Automotive Accessory (Part 4)

If you recall from the last part of the saga (Chapter 3), I had sort of a chicken-and-egg situation cooking with potential licensees for the new improvement patent. I also had a time limit to exercise an option to acquire the rights to the new patent. Everything had to click (or cluck) just right.

Company B didn't want to do a deal unless Company A did. But Company A wanted an exclusive license. So I started with the company that seemed to be more receptive and interested in taking a lead, Company A. (Remember, the president of this company initially escorted me out of his trade show booth, ready to give me the boot.)

It was time to get into serious negotiations with Company A. Although it was interested in an exclusive license, there was no indication of what minimum royalties it was willing to pay. A couple of weeks later it came back with an answer, and the amount was much less than I had anticipated. I had figured the royalty potential was around $25,000 per year. Company A's offer was in that range, not per year, but as a total figure.

With twelve years left on the patent rights, I believed it would be better to go with non-exclusive licenses, and I did. Company A was clearly disappointed when it learned that I would not offer an exclusive. Up to this point it had been enthusiastic about my mirror. Even so, it was still willing to license, though my mirror became just another item in its product line.

Company A wanted its expensive big-city law firm to negotiate the specific terms of the license. We had already agreed on a 5 percent royalty based on its selling price minus returns and trade discounts. This was fairly standard for this industry. Royalty payments would be generated quarterly along with a computerized statement of month-by-month sales. I would also receive twelve samples upon my request each year. This would allow me to keep up on the quality control.

Rather than have the company's law firm draw up the initial contract, I offered to initiate the contract and send it to the company's law firm for review. Even though I incurred a greater cost in doing this, it enabled me to originate the basic terms and provisions as a starting point for negotiations. I defined the invention not only as the patented product, but extended this to include know-how, trade secrets, improvements, and patents thereof. If I were to come up with new products along this same line, and if I later received design or other patents on improvements, this agreement would stand in force. I intended to make improvements to the mirror and possibly file for additional patents associated with them.

Since this was fairly standard language for this type of agreement, my terms were not questioned, and only minor changes were made to the agreement. I asked for quarterly inspection at the company's facilities, and it wanted to change this to annually. I agreed.

I also wanted to be able to inspect the company's records at random, and it insisted that any inspection of records be done through a CPA. This point was well taken, and I agreed to it also.

These were some of the minor points during the negotiation, and they went smoothly. Including mailing time, the whole process took about three weeks, and the president and I signed the contract. I had simultaneously exercised the option to purchase the patent rights from my former partner so there was little gap in cash flow between when money came in from licensees and when it went out to pay for the patent.

This initial contract was extremely important in setting the stage for my future dealings. It set the royalty rate, which became parity throughout the industry. No single company would want to pay more royalties than another. Also, since one major creditable company entered into a license with me, it made it more comfortable for others to follow suit.

Next we go to Company B. It had major accounts with mass merchandisers and did a high volume of sales with my mirror. It expressed an interest in licensing as long as other major companies were licensing at the same rate. Once it knew that Company A had signed a license, it wanted a copy of the agreement. I gave company B a copy of the license agreement after deleting Company A's name from it. It is a common courtesy not to share details that connect a company's name and its confidential contract terms with other companies.

One of the main things that Company B was looking for in the agreement was the royalty terms. It wanted to be sure that it wasn't paying more in royalties than anyone else. It also satisfied the company to know that I had a license agreement with a peer in the industry.

I basically used the same format for an agreement with Company B as I did for Company A, except for a few changes that would gain me more favorable terms. Most of these were accepted with little fanfare, and within a couple of weeks Company B returned a signed copy of the agreement without a review by its attorney.

I was also able to negotiate up-front payments from both Company A and Company B. This would help me negotiate favorable deals with other licensees. The initial payments varied from $100 to $1,500 or more depending on the company's current volume and past sales of my mirror. Interestingly, the amount of my initial payment wasn't always a result of the company's standing in the market; rather it was a function of how aggressively the decision maker negotiated. It all evened out in the end.

Seems like things were going smoothly, doesn't it? Well, up until then they were. Little did I know what was just ahead.

(To be continued in Chapter 7. . .)

HINDSIGHT LESSONS

These negotiations took a lot of communication, and 99 percent of it was done over the telephone. Each company was in a different part of the country. I would have been bouncing back and forth like a ping-pong ball had I talked to each representative in person. Not only would it have taken a lot longer, the travel expense would have exceeded the initial fees, plus some.

Treating licensees equally where royalties were concerned was an extremely important element in concluding these deals. It was essential to most of the companies I dealt with that they all pay the same royalty rates. No one wanted to pay more royalty than anyone else. Most of these companies try to position their products in major mass merchandising outlets. The sales of one item to one store can make a difference in the company's gross annual income to the tune of $100,000 to $500,000 per year. A difference in price of one cent can make a difference in whether a company gets the deal or not.

Therefore, if one company pays a 5 percent royalty and another company pays only 2 percent, it could ultimately make a huge difference in the success of the two companies' product lines. Paying me the same royalty established an industry standard for this product.

The will to persevere is often the difference between failure and success.

—DAVID SARNOFF

CHAPTER SEVEN

VICTORY THROUGH TEAMWORK

When to Use Professional Help and How to Find the Best

PEOPLE WHO DON'T TAKE RISKS GENERALLY MAKE ABOUT TWO BIG MISTAKES A YEAR.

PEOPLE WHO DO TAKE RISKS GENERALLY MAKE ABOUT TWO BIG MISTAKES A YEAR.

—Peter Drucker

Enlisting the help and collaboration of other professionals will greatly enhance your chances for success. These professionals may include patent attorneys, patent agents, and invention development agents.

The commercialization process must be customized for each invention. It requires an intense amount of personal communication with all of the parties involved. The professionals who are helping you must communicate openly with you, with potential licensees and other business contacts, and with each other. It is important to work with people you like and trust and to make sure that everyone on your team understands your goals.

Help with Patents

The patent strategy is a gamble. You do not want to spend any more money than is necessary to get the basic protection that you need under the law, yet you don't want to be so skimpy with your use of experts that you end up not getting the proprietary protection that you need. Once a patent attorney I hired failed to file a

Fire engine by N. Pierce (U.S. Patent No. 6394X), 1831.

patent application by the date required to meet certain international patent deadlines. As a result, we lost the opportunity to obtain patent rights in several countries that were important markets for us.

In another case, I was trying to save money by applying for a patent on my own and later found out that had I used different forms and a different process at one point in my application, I could have saved several hundred dollars and gained a few months of leeway. There's a lot to be said for the old adage, "People who represent themselves in law have a fool for an attorney and a fool for a client."

People who represent themselves in law have a fool for an attorney and a fool for a client.

Many strategies to save money on patenting are associated with some risk. These are complicated issues, and every situation is different. I advise people to talk with at least two or three patent attorneys or agents, plus an expert in marketing and commercializing inventions, prior to initiating the patent process. Many patent attorneys and agents will consult with you for the first half hour or so at no charge. If it's free, you have little to lose.

Patent professionals can also help you negotiate business contracts that will hold up in a court of law and provide any litigation services you might need. Selection criteria for these professionals are a little different.

I have used many different patent attorneys over the years, and I have used different patent attorneys for different things. I select them according to my need. When I have serious license negotiations or litigation to pursue, I use an attorney with experience in that area.

A patent attorney who has worked in corporations will have a better understanding of corporate protocol and corporate negotiations. Some patent attorneys were previously examiners in the Patent Office and specialize in applying for patents and prosecuting them, but they have not litigated patent infringement cases.

Successful patent attorneys will normally tell you what patents they have had issued for clients. They should also share the results of lawsuits they were involved in and that are a part of the public record. You can see exactly what kind of results a particular attorney has achieved.

Pick someone you like and can get along with. You should feel good about your communication with your patent attorney. You might try the attorney out by having them advise you on a certain aspect of your project.

A good patent attorney is not afraid to advise you to do a little market research before starting the patent process. However, like in any field, there are unscrupulous patent attorneys who will take on nearly any inventor as a client, claiming that inventors need patent protection to succeed with their invention. There is no doubt that patent protection may be an important element with your invention project; however, it is not necessarily the most appropriate first step.

Help with Commercializing Your Invention

When it comes to getting the best proprietary protection, there is often no replacement for the advice of good patent counsel. However, although patent attorneys and agents understand the intricacies of patent claims, they don't always understand the repercussions of those claims relative to existing market factors. You may also need to enlist the help of experts in invention marketing in conjunction with your patent expert.

Invention development companies typically offer evaluation services; that is, they'll assess your invention's chances for successful commercialization based on their expertise or research in your industry. They may also act as your agent in qualifying, submitting to, and negotiating with potential licensees.

Several years ago, I received a grant to study invention development companies in the United States. I was surprised to learn how few companies offer complete commercialization services for inventors. Of those legitimate organizations that I contacted (and there are, unfortunately, many illegitimate, rip-off outfits), most of their brochures presented them as offering complete commercialization services for inventors when in fact they may only offer evaluation services.

WORDS OF WISDOM

Be careful about falling into the trap of pricing attorneys strictly by their hourly rate. I work with a patent attorney who charges $250 per hour. He has over forty years' experience in the profession. He's top notch. Yet it costs me little more to have him prosecute a patent than it does for other patent attorneys with a much lower hourly rate. The reason? His experience saves his time and his clients' money. I end up getting a much better value for the dollar I spend.

Still, of those companies I surveyed that operate a full-service commercialization program for inventors, none had results worth writing home about. It would appear to be the nature of our business. The number of inventions successfully commercialized by the oldest and most prominent invention development companies, which hovers at 1 to 4 percent for those evaluated, is not much better than the rip-off organizations. Every now and then someone will advertise a 25 percent success rate, but this could refer to a lucky break on a couple of inventions. The more important question to ask is whether the company has stood the test of time.

Work with people you like and trust.

The company's success rate, incidentally, does not mean that any of these inventions ever made the inventor rich. It does not even suggest that the inventors made any profit on their inventions. It just means that an invention formally made it to the market, though for how long we don't know.

WORDS OF WISDOM

Commercialization services generally charge a fee and/or a commission of up to 60 percent. In other words, they receive a commission on your revenues. If they negotiate a deal that pays you $100,000 and a 5 percent royalty, a 50 percent commission to the commercialization service would earn them $50,000 and a 2.5 percent royalty.

I found only three services in North America that offer commercialization services for independent inventors for a reasonable fee while charging commissions in the 5 to 20 percent range.

RIP-OFFS

So what makes an organization a rip-off or scam? The basic distinction is in the fees charged for commercialization services. Rip-off companies charge thousands of dollars for their commercialization services, but they may only produce a boilerplate evaluation and send out form letters to companies they identified from databases. Generally, to be successful, this is not enough. At least you don't have to worry about having one of these high-pressure submission companies steal your invention. They're making so much money from inventors that they don't dare risk their reputation by stealing someone's idea.

These marketing companies have such high profits that they can list toll-free numbers in phone books nationally, advertise on TV, present polished brochures and promotional materials, and call you several times to encourage you to use their services. I often see inventors who wish they had never used those services and spent all that money.

How do these companies operate? Typically they start out charging less than a thousand dollars for a preliminary report, which turns out to be full of useless statistical information and a lot of ego stroking. Some of these companies actually provide you with some really good information about patenting and invention development. This gives you a bit of a head fake.

In their next step, the companies charge $7,000 to $25,000 and guarantee a patent and national or even international exposure. What you don't find out until it's too late is that in many cases only a design patent is obtained, which is very limited in its scope of protection. Also, the potential licensees that they submit the invention to are not carefully targeted; you rarely end up reaching appropriate potential licensees.

Reputable commercialization firms charge a reasonable fee and do real research.

Typically, they give your invention minimal exposure at national trade shows. A large part of their efforts at the trade shows is to promote their own company, and your invention will be displayed in a book with as many as five hundred other inventions. Very few manufacturers pay serious attention to these types of promotional efforts. The marketing company has, though, fulfilled its contract.

Inventors' groups around the country frequently do battle with these marketing services. Get in touch with your local inventors' group and ask which companies their members complain about and which they appreciate. This is the most reliable method of learning about good and bad invention marketing companies.

Complaints against invention marketing companies are common enough that the U.S. government has become involved. The USPTO's Office for Independent Inventors now handles this area. The Resources section tells you how to reach them.

The Boston office of the Federal Trade Commission (FTC) played a significant role in obtaining court-ordered consent decrees from at least two major invention marketing companies. The court has ordered these companies to put over a million dollars in escrow to repay inventors who have made claims and allegations against them. If you are not happy with the service you get from an invention marketing company, you should contact the FTC in Washington, D.C. There may be money available to you if you choose to make a claim. If you have an attorney pressure a major invention marketing company, it may give you a full refund, but then you still have to pay your attorney.

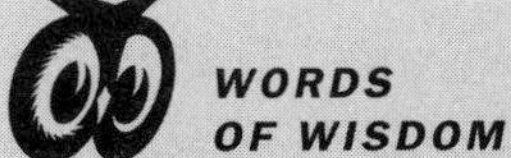

The rip-off's cost to prepare a boilerplate evaluation, file a design patent, send form letters, and include your invention at a trade show booth is probably under $1,000. If a company charged you $12,000, where'd the other $11,000 go? You can see why these companies can afford to advertise.

Why do these notorious rip-off companies still flourish? Most government employees or national inventors' group members have never successfully commercialized an invention. Although they may be able to recognize the good guys from the bad guys, they are hesitant to point fingers due to liability reasons. They'll supply an inventor with a list of questions to ask a potential invention marketing company; however, often these questions are so generic or inappropriate that inventors aren't able to successfully distinguish between a company that's really going to help and a company that's just going to take their money. Because the rip-off companies will try to mirror ethical companies' fees, published success rates, disclaimers, or company structures (such as sliding scales or tiered risk programs), it's often impossible to tell the difference between the two just by asking a few questions. This is one reason why these alleged rip-off companies flourish.

So how, then, do you find an ethical company that will really help? Reading this book and educating yourself about the commercialization process is a good first step. Network with other inventors and attend meetings at your local not-for-profit inventors' organization. Other inventors will candidly share information that a government agency or national inventors' group could never disclose. Plus, the information tends to be more timely. To find these valuable resources, refer to the Private Organizations listing in the Resources section.

USING PATENT ATTORNEYS TO COMMERCIALIZE INVENTIONS

Although some patent attorneys advertise marketing commercialization services for inventors, few if any offer an in-depth market evaluation or any commercialization services of substance. Patent attorneys tend to rely on the strength of your patent rights to commercialize your invention. With the patent in hand, they find companies listed in a database such as *The Thomas Register* or by using U.S. government SIC codes. Then they send form letters to these companies announcing your patent. This is not necessarily a bad service; however, I have found that it is hard to target the right companies and to get consistently good results using this method.

It can be hard to recognize any difference between the invention marketing methods used by patent attorneys and the methods used by unscrupulous invention development companies. The major distinguishing feature is that those inventors who have started with a patent attorney may at least have better proprietary protection, as useless as it may be, and it will only have cost them $4,000 to $10,000. In contrast, an alleged "rip-off" company's fee may range from $7,000 to $25,000. Sometimes this includes filing a patent application, and sometimes not. This is not to suggest that charging such a fee is inappropriate in all cases; it just depends on what you receive in the way of service.

Help with Patent Infringement

The burden is on you to take legal action against anyone you believe is infringing on your patent rights. The emphasis here is on the word *burden*. Let me explain.

WHEN SHOULD YOU SUE FOR INFRINGEMENT?

The short answer is, never, except under extremely exceptional situations. The simple fact is that it costs a whole lot more to sue someone than what you gain in 99.99 percent of all circumstances.

Count to a thousand—twice—before you decide to sue.

It is estimated that the median cost for one party in a patent infringement lawsuit is $2.8 million. This figure reduces to $300,000 per side when the lawsuit is worth less than $1 million. The average time to conclude a lawsuit worth less than $100 million is four years. When more than $100 million is at stake, the average length of the lawsuit is twelve years. Infringement litigation is normally not considered unless the market potential for the invention is worth millions of dollars. Sure, you can send a letter to someone and threaten a lawsuit; however, a knowledgeable attorney will recognize that your threat is useless without the financial justification behind it. This is yet another reason why I put particular emphasis on orchestrating win-win deals to begin with.

Even if an infringer were to take your threat of lawsuit seriously, the infringing company would likely not be liable for damages for infringement until after it has been put on notice, and after it has satisfied itself that its product does indeed infringe your patent. The burden of responsibility is on you to put the infringing company on notice. So in other words, until the infringing company knows that it is indeed infringing, it can get away with murder. Since patents are considered defensive tools, there is not much you can do about this.

When considering whether a company is actually infringing on your patent, you must first determine whether it truly is infringing on your patent claims, and whether your patent is valid. Most issues of patent infringement are not black and white; there is a lot of gray area.

The first step may be to have an attorney render an opinion regarding the scope of your patent's claims relative to the features of the potential infringer's product. Is it possible that one of the key elements of your broadest patent claim is not included in the infringer's product? What are the key elements in your patent claim? Are they essential? To obtain an attorney's opinion regarding scope may cost between $3,000 and $15,000 plus. Whether or not you have a scope opinion rendered, the infringer very likely may. In most cases scope opinions ordered by the infringer miraculously come back benign. They'll claim that they are not infringing. Again, the burden of proof is on you (and your dime).

Blind persistence yields unpredictable results. Wise persistence yields desired results.

To determine whether your patent is indeed valid, a patent attorney will obtain your entire patent file docket (or "wrapper") from the USPTO. These are usually at least a foot thick. Then they will perform a serious review of all of the prior art patents and literature related to your invention. This process may be more thorough than when your patent was originally prosecuted. Costs for a validity opinion start between $5,000 and $20,000. If the other side's law firm believes that your patent is invalid, it can cite this under the law to have your patent disallowed, or at least to have certain patent claims disallowed.

There are two primary ways a company can do this. It can file a Declaratory Judgment Action suit in Federal District Court. The action may take three to six months minimum to try, cost the company $150,000 to initiate, and cost you a similar sum to defend. Or, it can make a request for reexamination in the USPTO. The cost for each party is similar to the cost you incurred to obtain your patent. During this process, you could request a reissue of your patent and actually add claims to it. On the other hand, if you accuse someone of infringing on your patent, you stand a chance of losing your current patent rights in the process. This gets touchy.

Thus far we have been discussing various maneuvers that could take place prior to initiating an actual lawsuit. If both parties each have a scope and validity performed, the cost could range from $10,000 to $100,000. If more than two parties are involved, it could be more. If the scope and validity findings are in your favor, prior to an all-out legal battle, you can request a preliminary injunction to stop the infringing while a lawsuit is undertaken. Plan to budget at least $150,000 to do this.

Who is going to pay for these expenses? The information found in the scope and validity is necessary to form an opinion as to whether a lawsuit is even practical, or whether it might be successful. As you can imagine, it is difficult to entice anyone to take a gamble on your cause prior to knowing whether you have a good case. This is why the only people you typically see in the news regarding patent infringement suits are very large companies and those involved in very large markets. Most cases also involve markets that are already established. After all, if the market is not already established, how do you know how much you are losing?

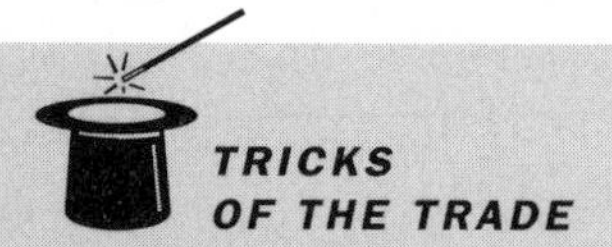

So, Do You Still Want to Sue the Bastards?—I once threatened to use a unique ploy, my secret weapon, against a company. It created such concern that the company conformed to my wishes without further ado. I think it can work for others, too. If a manufacturer that is selling your product or technology to the market is indeed infringing on your patent, it is not the only one infringing your patent; so are the distributors, retailers, and end users. After all, a patent covers those who make, use, and sell the patented item.

I have been advised that in an infringement lawsuit, because of the *Doctrine of Exhaustion of Rights*, you can only seek damages from one level of distribution. You can seek damages from the manufacturer or the distributor or the retailer or the end user, but only from one of these levels. You cannot, for example, sue the manufacturer and then seek damages from the distributor to which that manufacturer sells. Normally the lawsuit is filed against the manufacturer because it has larger pockets to dip into, and because its sales contracts usually indemnify the retailers and distributors from such infringement lawsuits. You could, however, opt to sue distributors, retailers, or for that matter, the end user. The very threat to do so, my secret weapon, could stop a manufacturer's infringement cold.

The last thing that a manufacturer wants is problems with its customers, i.e., the distributors and retailers. Retailers and distributors certainly do not want to be messing around with a patent infringement case even if someone else is footing the bill.

A lawsuit could affect a manufacturer's relationship with a distributor or retailer, possibly the manufacturer's entire account with them, and their public image. Such a quagmire could affect the manufacturer's sales well beyond the income derived from your specific invention. This could significantly add leverage to your ability to influence the manufacturer. This is a good example of why it is important to have a firm understanding of your market and the channels of distribution within your market.

Be aware that such a strategy could backfire. You may win the infringement issue, but in the process some of your key distribution outlets may never want to carry your product or any future products from you. Then again, you may not have anything to lose from making the threat. This is a tough call and requires serious soul searching.

PATENT INSURANCE

In the Resources section of this book, you'll find the Intellectual Property Insurance Services Corporation (IPISC) of Louisville, Kentucky. IPISC is one of the nation's larger insurer of patents. For fees running $1,000 and up per year, IPISC will insure your patent against potential infringement for $100,000 to $3,000,000 and more. The specific price varies depending on the risk and the level of protection.

If a company infringes on your patent after you have obtained the insurance, IPISC will help pay for litigation expenses up to the amount insured. Although your insurance coverage may be less than the cost of the average patent infringement case, it is rare for a lawsuit to reach the stage where big bucks would be expended on a major trial; in other words, most cases are settled out of court or never get to that point.

IPISC may have the scope and validity opinion rendered for your invention to determine whether there is a firm basis for a lawsuit. This alone can be enough to stop a lawsuit and force negotiation. When an infringer knows a patent infringement insurance company is behind you, there is a greater chance that the infringer will be willing to settle with you out of court; the smaller the infringing company, the greater this will affect them.

This all sounds fine and good, but is there risk involved? The primary downside that I have seen is the cost of the insurance policy itself. The cost for adequate insurance is more than what most inventors may make from royalties on their invention. An average license deal will take at least three years before you start to see royalty revenue. So you will have substantial costs for insurance before you know if it's really needed.

Another risk with patent infringement insurance is that it is normally only valid for infringement that occurs after you have obtained the policy. It does you no good if an infringer started selling its product, in a small test market for example, prior to when you first obtained your policy.

For example, a manufacturer with an infringing product may offer it to a few select customers to try out and see whether they like it. After receiving positive responses from customers, the company decides to invest in tooling to produce the product. The next year, it finishes the engineering and tooling and starts to crank out the product. It may first offer the product to certain distributors to test market on a small scale. The manufacturer may decide to not mass produce the product until after the test market to be sure that the product will actually fly. After it discovers that the product has significant potential, it starts to contact the buyers of major mass merchandisers or other mainstream retailers. The buyers agree to place the product, but it can take another year before you see the product on the shelves. Three to five years may go by before you realize that there is an infringing product on the market, yet the first actual sale took place long ago. In this event your insurance coverage, bound after the initial test sales, may be void for that infringer.

I know that this scenario happens because IPISC has hired me to perform market investigation for infringing inventions, and I have sometimes found the first date of sale to be years before anyone suspected.

CONCLUSION

As your invention project matures, there is a greater likelihood that you will want to seek the help of professionals. The questions are who and when? *Who* depends on your specific needs. *When* will be dictated by your budget as well as your needs.

If you have the gift for gab and want to save money you can do much of the footwork yourself. But when you need to bring in professionals, bring them in early and familiarize them with the project immediately. Then make periodic phone calls when you have questions, concerns, or even good news. This keeps your "partner" informed and enthused and allows you to control the process (and your pocketbook).

When should you consult with a patent agent?

- At the very beginning of your project, before disclosing your invention to others.
- When you are ready to perform a prior art search.
- To discuss the patentability opinion.
- When you are ready to file a patent application.

When should you consult with a patent attorney? For all of the same cases you would consult a patent agent, and:

- When you need a customized disclosure agreement.
- Whenever someone presents you with a legal document.
- When the other party's attorney contacts you.
- When you are a partner in an invention, an employee, or a grantee.
- When you are ready to undertake a formal business relationship.
- Prior to and during license negotiation.
- Before signing any document.
- When considering foreign patents and contracts.

When should you consult with a licensing or marketing agent or invention broker?

- After your invention is documented, witnessed, and dated.
- After you receive the patent search results.
- Prior to filing a patent application.
- Prior to revealing your invention to companies.

- Prior to any public disclosure.
- After you have done basic market research on your own (unless you hired the agent or broker to do it).
- Prior to or when you need help with negotiation.
- When you are trying to determine a commercialization strategy.
- When you are totally confused and before throwing in the towel.

THINGS TO REMEMBER AND CONSIDER

- Everything and everyone costs money. Figure out what you need, identify the value it will bring, and buy (or hire) no more than necessary.
- Approach any deal with vulnerability—but don't show it.
- Turn failures into new opportunities; build on what you have learned.
- Celebrate success.

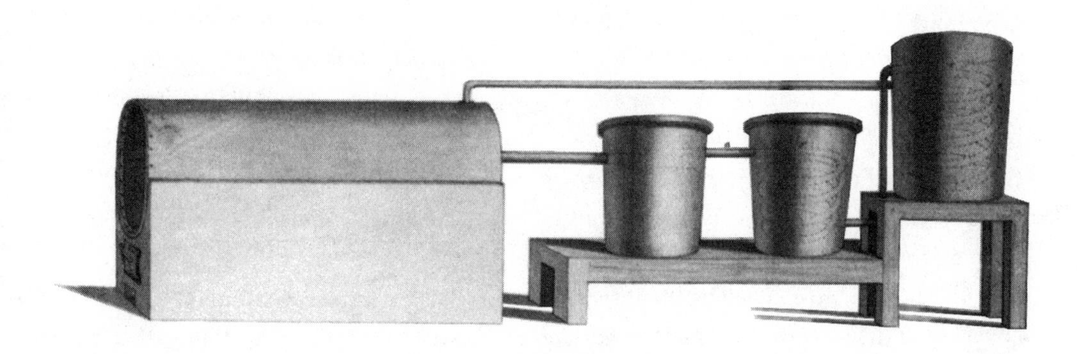

Distilling by Thomas Lawes et al (U.S. Patent No. 5175X), Louisville, Kentucky, 1828.

CASE STUDY

An Automotive Accessory (Part 5)

In the previous episode of the Docie wedge mirror saga, Companies A and B signed license agreements for my newly acquired patent. This was a good start, but no single company came close to controlling the market, so I had quite a few more to go. I would have to license at least a dozen more companies to capture most of this very large and fragmented market. Part of my decision to not go exclusive with Company A arose from the fact that I would have missed out on most of the market. I was starting with a marginal patent position, favorable test market results, and a whole lot of people to deal with, each coming from different perspectives. How would I best leverage this tenuous position?

My patent attorney advised me that the mirrors already on the market did indeed infringe on my patent, but that we had about a 50/50 chance of holding up in court. Furthermore, it would be very possible for someone to design around the claims of my patent so as not to infringe. I still decided to play hardball, hoping I could convince the infringing companies to license with me in order to avoid an infringement suit. (This was something of a bluff, because I did not have the financial resources to go through with a patent infringement case.) I sent a formal opinion letter prepared by my attorney to the potentially infringing companies accusing them of patent infringement. Then I followed up with a phone call to the various presidents stating that I really didn't want to take legal action if I could help it. I would rather create a win-win situation. The phone calls were a relief to the company presidents because licensing with me would save them money and avoid the hassle of dealing with lawsuits. Luckily, most of them were pragmatic about the situation and appreciated my willingness to be reasonable and negotiate. This is another example of the good guy-bad guy routine working well. My patent attorney played the bad guy, ready to pounce, and I portrayed the good guy, ready to mitigate. In our high-stakes role playing, I wanted to be the good guy because I would be the one to do the subsequent negotiations. It worked!

Of the dozen or so companies that I sent the opinion letter to, about half were doing a low volume with their product, and they decided to drop it from their product line rather than mess with a license. A couple of the companies did decide to sign license agreements. That left me with the three companies I came to call "the Bermuda triangle of the Midwest."

These three companies, all of which I hoped would sign license agreements because of their decent mirror sales, all wanted to be sure that each of the two others were committed before they signed. The threat of the patent attorney's opinion letter did nothing but create a stalemate. I had to jumpstart things.

Meanwhile, I had performed a test market at local stores to determine consumer preference for style, shape, size, and so on (see "Test Marketing" in Chapter 3). Armed with favorable test market results for my new designs, I called the president of one of the companies (Company C) and offered to forgive the company for any past-due royalties and liabilities associated with patent infringement in consideration for the company paying a 5 percent royalty on my promising new designs and improvements. Here is how I approached the presidents. I said, "I just concluded a market test and found that my new mirror designs are preferred by 78 percent of consumers. Plus I learned more about consumer preferences not previously known, the results of which I am willing to share with you if we can work a deal. I am so excited about the future prospects from these test results that I am willing to ink a deal based primarily on the improvements and forgive you of all past sales and liability."

This completely changed the tone of our negotiations, and I was met with an amiable response.

Rather than send the company a long, formal license agreement, I created a one-page, letter-style agreement (see following page). I forgave them for past-due royalties, and they agreed to pay the 5 percent royalty on a quarterly basis, not only for the patent rights but also for improvements, trade secrets, know-how, and in particular, the new designs. The duration of the agreement was left open forever. I left a space for the president to approve these terms and return it to me. Since I was forgiving past-due royalties, and with the inclusion of test-market data to help them get started, the company was willing to pay up-front fees. The fees that I proposed were largely commensurate with the amount of infringement that had taken place in the past by the company, based on the 5 percent I otherwise would have received. By changing my approach I allowed the company to save face and not have to admit to past infringement, and I still got everything I wanted and more. If I had submitted a twenty-page license agreement, it surely would have come under much scrutiny and a tremendous amount of negotiation with their attorney; I doubt the deal would have ultimately succeeded.

The next two companies (Companies D and E) were harder to reel in because they definitely did not want to sign deals unless all three agreed to do so. Company C had agreed to the terms of my one-page license agreement, but had not signed it yet and wouldn't unless they knew that the other two companies also signed an equivalent deal. So I went to Company D and explained that Company C had already agreed to enter into the license and asked if they would be willing to sign if Company E did too. They indicated that they would.

I went to Company E with the same enthusiasm over my test-market findings and indicated that Companies C and D had tentatively agreed to enter into a license agreement and asked Company E to do the same. In light of this, the company agreed. Then it was back to Company D. When I told Company D I had (verbal) agreements with Companies C and E they were ready to deal as well.

Here is a copy of the letter I sent to Company C.

January 19, 1988

RE: DOCIE BLIND SPOT MIRROR TECHNOLOGY
Confidential Information

Dear Andy,

As per our conversation January 18, 1988, this will confirm our understanding.

Ron Docie will not take legal action against Company C for past actions regarding the wedge blind spot mirror technology, U.S. patent no. 4,311,363.

Company C agrees to review improvements and modifications to the existing blind spot mirror technology on a confidential basis and to not release said information without the express written permission of DOCIE. COMPANY C is under no obligation to accept DOCIE's blind spot mirror technology. This paragraph does not apply to information already known by and documented by COMPANY C.

In the event COMPANY C uses or commercially exploits DOCIE's designs and other trade secrets, COMPANY C shall purchase only from DOCIE's approved sources or pay DOCIE a five percent (5%) royalty on blind spot mirror revenues generated (COMPANY C's selling price minus actual returns, normal trade discounts, and sales commissions to a maximum of five percent (5%)) and paid quarterly.

DOCIE shall provide COMPANY C with complete results from the Blind Spot Mirror Survey and design improvements, within thirty (30) days. These improvements will offer a better viewing area than existing products on the market. The Blind Spot Mirror Survey results show greater consumer acceptance of DOCIE's designs versus existing designs. The survey also shows preference for shape, size, model, border, border size, viewing angles, and more.

Also, COMPANY C shall pay DOCIE an initial fee of $1,500 due immediately.

Sincerely,

Ron Docie

Enclosures: Invoice
cc: John L. Gray, Esq.

Here is the letter of response from Company C.

February 2, 1988

Mr. Ron Docie

Confidential

Dear Ron:

Enclosed is our check for $1,500.00. This check covers our understanding and License Agreement as per your letter of January 19, 1988.

We look forward to receiving the Blind Spot Mirror Survey and the design improvements. Hopefully, we can work closely together in the future.

Very truly yours,

President, Company C

AJR/eab

Enclosure

(*Footnote:* Company C ended up manufacturing several of my new designs.)

When I went back to Company E and explained that Companies C and D had committed to the deal, they said they would also. I told Companies D and E that I would send them letters that spelled out the terms of the deal that Company C had agreed to. Since the contract came in the form of a letter, they were more willing to close the deal without the benefit of legal counsel, and they did sign the contracts. Now the triangle was complete.

In the following twelve-month period, I negotiated twelve additional licenses (with varying formats depending on the situation), all with advance payments. That's an average of one per month. This kept me pretty busy, as you might expect.

My next task was to get these companies to take on my new line of mirrors. Company C was interested in pursuing my new line of mirrors in order to get a jump on the other companies and also to be the first to offer unique products in its spot mirror product line.

Although it had a non-exclusive license with me, it appeared to be the only company that would initially tool up my new product line. Since they made a commitment to pay for the tooling, which totaled nearly $50,000, I agreed not to aggressively pursue the commercialization of my new designs with other companies. My agreement to Company C was made under the good faith that they would aggressively pursue this virgin market.

But a year went by, and Company C still did not have any new products from my designs to offer at the annual trade show. This was very disappointing. At one point they had to switch mirror manufacturers and start from scratch. It was a large company and tended to move slowly, and I was losing out because of their sluggishness. I felt they breached the good faith element, and there was no indication that things were going to progress any better.

So I gave up on Company C and approached two of my other licensees to ask whether they would be willing to tool up my new product line, and, by the time the next annual trade show came around, Company B had some of my new designs available at their exhibit booth. Coincidentally, Company C finally got their act together and displayed my entire new product line at its booth also. Company C was extremely upset that they were not the only manufacturer with my new products—they felt betrayed. Regardless, in hindsight, I believe I was appropriate to go to Company B. It was a tough moral decision, but Company C did not live up to its promises. As it turned out, Company B was able to penetrate the market with some of my new products, which was reflected in handsome royalty payments. To this day, Company C is still floundering with its product line. Because of the popularity of my new designs, three of my other major licensees were offering my new product line soon after Company B.

From the time I acquired possession of a U.S. patent in 1987, it took four years for this licensing process to be completed. The royalties nearly doubled between 1991 and

1992. Royalties did not exceed $15,000 per year until 1992. They have increased every year since. All the major players in my industry are now paying me ongoing royalties.

Photo © 2001 by Sam Girton

This photo shows the sixth (or so) generation of the Docie wedge mirror design. Each licensee features its own unique styles and packaging. The product specifications and sizes also vary between licensees. The common element is the wedge-angle, regardless of whether the mirror is rectangular or round. This is an example of how a utility patent is better than a design patent because you are not limited to style and size unless specified.

Over time some of my licensees have changed their product line emphasis and decided to get out of the mirror business. A few new companies have started up, and I have negotiated licenses with them. There are a couple of companies that were holdouts, and it took five to six years to finally conclude a deal with them. I had to be patient and hound them every year until the guilt was too much for them. If I had not gotten in their face every year at the annual trade show and shaken their hands, it would never have happened. Who said licensing is not hard work? It was the personal element that made the difference.

The 1995 annual APAA automotive trade show in Las Vegas was the first show in my twenty years' experience with the Docie mirror where I found no infringers, and every manufacturer that had my mirrors (about fifteen) was under a payment plan with me.

One might wonder why a dozen or more companies would be willing to pay royalties on an ongoing basis when it is clear that they could either design around my patent, conspire with others to avoid paying royalties, or simply deny that my patent has any validity. I am sure that they all realized that it wouldn't be much of a legal battle if I were forced to sue each of them in their respective states. My legal costs would be astronomical. So why are they continuing with our deals?

There are several reasons for this. First, it is not worth their time and money to do any type of legal battle for such an item. Most companies would just as soon drop the product if it came to that.

Second, I brought added value to the product line. I had agreed to keep them informed about consumer preferences for size, shape, and so on, and to monitor changes in mirrors. This is something that none of them cared to do on their own.

Third, I keep up with other mirror inventions; from time to time my licensees send me a mirror disclosed to them by another inventor and seek my opinion as to whether the invention has merit, and to determine its pros and cons. Last and not least, I have been a diligent licensor. I have maintained communications with the presidents of these companies over a period of years and have sent follow-up letters to verify any understandings that were made verbally on the phone.

At times it may have seemed like I was hounding them; however, I believe that most of these executives appreciated that my efforts were aggressive without being too overbearing. They knew firsthand that I worked hard for my royalties. My files on this one invention alone are nearly five feet thick with notes of conversations, copies of correspondences, and collateral material.

Another advantage for my licensees is that they can now place a patent number on their products and should they ever decide, they will be prepared to introduce new products along these lines. Although each licensee knows that there are other licensees they would have to compete against, they will at least know the limit of that competition. At this point, others that try to infringe would be coming up against at least six substantial companies with licenses on my invention. This helps to keep out the players that might cause problems.

HINDSIGHT LESSONS

In rounding up the companies infringing on my patent, I found that a carrot worked better than a stick. After having threatened some of my potential licensees, I found myself in a standoff. It certainly made no sense for either of us to sue the other. It was not until I responded in a very positive manner by suggesting that I would forgive transgressions and continue with an improved product that progress was made. Since my potential licensees really didn't know whether I would sue them or not, it was a bit of a relief for me to agree to forgive past-due royalties in exchange for their entering into a deal that would instead guarantee future royalties.

The purposes of the up-front fees were to help cover the cost of my market research and to provide companies with the benefits of my findings. I suggested that these research results might help them to eventually improve their sales. In their minds, this is really what they were paying for. In actuality, what I did was figure initial payments on a sliding scale proportionate to what each of these companies had infringed in the past. So essentially they were reimbursing me for past-due royalties, but we called it something different so they would be more comfortable about paying an initial fee. They were happy and I was happy.

This is an important distinction because the companies would have set an unfavorable precedent for themselves if they had agreed to pay past-due royalties. They did not feel that they were obliged to do so, and they did not want to open the door for other inventors to seek past-due damages in the future. Renaming and rejustifying the reason for the initial payment helped them to save face, avoid potential future liability, and at the same time enter the deal.

Going with the flow was essential. Had I developed a scientific formula for payments, it would have met with failure because the companies already had an idea of what they wanted to pay. Had I insisted that the big companies pay more initial fees and the smaller companies less, this too would have failed. Some larger companies paid less, and smaller ones, more. As it turned out, I received about the same amount of up-front fees as I would have had I tried to institute parity on this point.

However, different companies have different mentalities about honoring intellectual property. For example, a computer company may have a hundred engineers working to solve problems, and an inventor comes to the company with a solution to one of the problems. The inventor requests a royalty that is a hundred times more than the reimbursement one of the engineers would receive. Sometimes it is hard for either party to understand the perspective of the other; this results in another standoff. Understanding the other party's perspective, therefore, is an essential element in successful negotiation. This goes a long way toward establishing terms that both parties would ultimately accept.

It is very important for me as a licensor to maintain diligent policing of the market to make sure that clandestine companies are not trying to sneak in nonlicensed products. As you might imagine, a company or two doing this could mess up my entire royalty income from my current licensees. Here again, setting a precedent is what is important. My concern isn't whether I lose royalties from a small operator that chooses not to pay them. It is the larger royalties I would lose from my major licensees if they thought that I was not protecting their interests.

There are also legal reasons for treating people fairly and equally: federal antitrust laws regulate the type of fees that can be charged in industry. If you allow one licensee to operate under a different set of rules than another, you may be violating the Fair Trade Practices Act. The interpretation of the laws regarding this change from time to time, so it is important to consult with a patent attorney who is familiar with these antitrust laws when setting up licensing contracts, setting prices and policies for distribution, and orchestrating other such transactions. Every now and then we hear about a major lawsuit against a company because of this, and you certainly wouldn't want to be in that position if you could help it.

Patience played an important role in my project. Although I certainly wanted to get rich quick, I was realistic when I learned that projected time frames did not meet my original expectations. If you are going to take an invention into the national and international markets, it is a long-term proposition. There are, of course, exceptions. Sometimes inventors luck out, and a major company grabs their invention for a million dollars, but this is incredibly rare.

Reasonableness also supported my efforts. Had I approached this venture with the attitude that my invention was worth a million dollars, take it or leave it, I would have been left in the dust. There are more inventors left in the dust than not.

The Docie wedge mirror has now had a market lifetime of over twenty years. Tenacity and persistence were major factors that kept this project alive. Some inventors carry this too far. There is a difference between being optimistic and being unrealistic. In hindsight, if you consider all the hours that I put into this project, my ultimate payment was not outrageously high. Additionally, I was able to use the skills I learned on this project to help other inventors and to make a good business doing so. In this respect, it was a paid education.

Willingness to admit to being wrong was a recurrent theme in my experiences. I now believe I was wrong in deciding not to have the mirror initially manufactured overseas. My first selection of a patent attorney was a mistake. Before Rubber Queen came along, I had a chance to sell out. That money, if well invested, would have made me more money than all my eventual royalties. Come to think of it, I erred a lot. I guess we can count how many times we were thrown off the horse, or we can count the number of times we got back on. I kept getting back on. It was certainly a rough ride; however, I am now the proud owner of a successful invention, which was my goal.

There are two mistakes one can make along the road to truth—
not going all the way, and not starting.

—BUDDHA

CONCLUSION

THE END OF THE RAINBOW

As I reflect on my career in the invention business, I recognize certain patterns: business was always unpredictable, transactions took a lot longer than expected most of the time, and proposed budgets were exceeded all of the time. Financial rewards were rarely as high as I dreamed, and invention failures happened a hundred times more often than not. Rewards have not always come from the expected, and financial gain has only been part of the equation.

If you think about how you spend your time, very little of it is actually spent walking to the bank with a bag of cash. Except for the time we spend with family, community, and—ahh!—on vacation, we are in the trenches. For some of you that means keeping your nose to the grindstone in a laboratory, inventing away. An apt reward for you may be the flexibility to do this more often or with more choices. For others it may be seeing your ideas in the spotlight, in the marketplace, knowing you were the original conceiver. Many inventors experience this, but not necessarily with financial awards attached.

And then there are marketers like me. My intangible rewards have been many. For example, when I go to a major industry trade show, it's almost like a vacation from the office and a (business) family reunion all wrapped up in one. I've built

Baking iron or gridiron by E. Skinner (U.S. Patent No. 6164X), Sandwich, New Hampshire, 1830.

friendships over the years with key players in industry. We can joke, laugh, and discuss the next crazy contraption I have in my bag. I can't think of a successful deal that didn't come with a new friendship attached. It's a complex network of people from different walks of life, a variety of locations, and industries aplenty, uniting with the common goal of doing good for others and getting a just reward in the process. There are gimmicky fads that turn a quick buck and make millionaires along the way. But that is a fast-paced, dog-eat-dog, here-today-gone-tomorrow kind of game. It is not an arena in which I choose to play.

I just invent, then wait until man comes around to needing what I've invented.

—RICHARD BUCKMINSTER FULLER

The more appealing things about the invention business are the intrigue, the suspense, the hope, and the dreams. We inventors comprise an optimist's club. Being in this business requires looking ahead at what can be, and not accepting something lesser when a better solution is possible. What's kept me here is the challenge of providing a solution that makes people's lives better and our environment more sustainable. We can consume less (in the past people consumed less and still had fulfilling lives) or we can make those items that we do consume better, more effective, and friendlier to the environment. Ideally we would do both. The latter scenario, consuming more prudently and designing the best possible products and technology, allows us to leave a nobler legacy—and it keeps inventors in business.

It takes a passionate inventor to have the tenacity and drive to overcome the obstacles to commercialization. But, remember, the obstacles are mostly the same whether you have a financial winner or not, which is the whole point of this book and the basic practice of my professional career. You need to learn where your invention fits into society, and do so as quickly and inexpensively as possible. If commercializing your invention doesn't meet your expectations, either financially or personally, get out and move on. You can be passionate about inventing and the possibility of getting rich; however, there are various ways of getting there. If in your heart you know that your invention is meaningful, revolutionary, and usable, then it probably will be successful; the question is when. Every day I meet inventors who are ahead of their time. And that's okay; the world will catch up to you.

If you are ahead of your time, stay on the cutting edge and be ready to seize the moment when it comes. To keep all of this in perspective, remember: The bad news is, change is slow; the good news is, change is slow.

AFTERWORD

The Future for Inventor's Agents

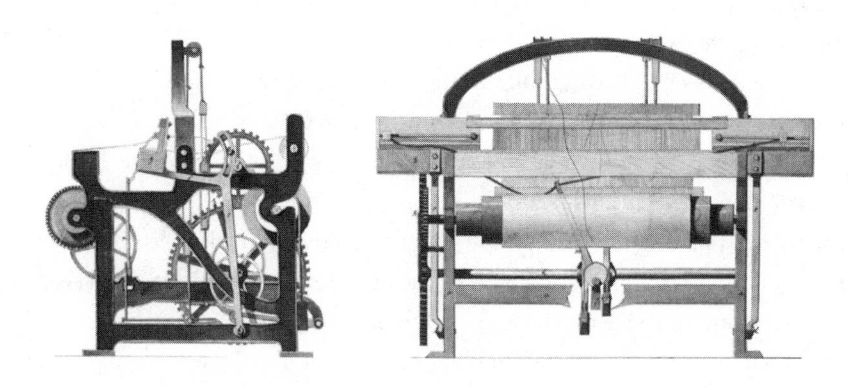

I hope there is a budding invention marketing agent or two reading this book. Sometimes it's like being the Maytag repairman; it gets lonely out there. That's because few professionals understand the invention marketing game well enough to stay in business for more than two years.

If you look at this profession critically, inventor's agents, or invention brokers, are really change agents charged with a fiduciary responsibility to society. Sounds heavy, doesn't it? But seriously, when you place a worthwhile invention into society, end users gain the most; the inventors only get a fraction of the reward, if that. Think of it like this. If your invention adds ten years to the average life span of its beneficiaries, how can your reimbursement compare?

I tirelessly represented a doctor who developed a possible cure for breast cancer. His discovery could be used to make vaccines and shortcut diagnoses by ten years, with no mammograms. His initial clinical test showed incredibly high sensitivity and specificity. This doctor, although a nice person in general, was the most unpredictable character I would ever want to deal with. I took his project where no one else said it could go: straight to the major pharmaceutical companies and right to

Loom by W. H. Howard (U.S. Patent No. 5826X), 1830.

the directors. The premier company with the greatest R&D resources and experience in the industry was jumping through hoops to learn more. It even broke precedence and offered us a very good confidential disclosure agreement. All we had to do was get to the first meeting, and even then we did not have to reveal trade secrets to get the ball rolling.

But my client refused to proceed because of fear, paranoia, and greed—emotions embraced by many inventors. In fact, I've experienced a lot of this from clients. If I were to discontinue projects because I realized the inventor was unreasonable, I'd be quitting for the wrong reason. On a higher plane, it's not the inventor that I work for; it's society. Why should I let immature personalities stand in the way of progress? Many times I've had to help a project along in spite of the inventor. This is a lot easier to do when you focus on the bigger picture.

This, and issues like it, raise interesting ethical dilemmas. Should I invest more in the breast cancer project even with the small chance that it will be successful? Is it worth the possibility of saving thousands of lives? What would you do? There is seldom a dull moment in this business.

The world is getting more and more complex, and at a quick pace. Each new challenge presents a problem, and with it, the need for a solution. It's open season for inventors, and likewise for the agents who help them. If you like the starving artist idea, and you like the idea of helping others, you like the notion that someday there may be a cool million in it for you, and you like meeting new people, then maybe this business is for you.

I'm always willing to consider new associates. I have developed a training and apprentice program for those interested in learning more. I know of no formal school that offers a degree in this specific profession. So by all means, write me a brief note if you are interested in becoming an inventor's agent. The world needs you.

APPENDIX A

General Information Concerning Patents

This section is a condensed version of a brochure published by the United States Patent and Trademark Office (USPTO or Office) giving a comprehensive overview of the operation and handling of patents. The full text of this overview can be obtained from the USPTO, or on the Internet at www.uspto.gov/patents/resources/general_info_concerning_patents.jsp.

Although the head of the USPTO has been referred to as *Commissioner* since 1836, recent legislation changed this title to *Under Secretary of Commerce for Intellectual Property and Director of the USPTO.*

About the U.S. Patent and Trademark Office

The USPTO is an agency of the U.S. Department of Commerce. The USPTO administers the patent laws as they relate to the granting of patents for inventions and performs other duties relating to patents. It examines applications for patents to determine whether the applicants are entitled to patents under the law and grants the patents when they are so entitled; it publishes issued patents and various publications concerning patents, records assignments of patents, maintains a search room for the public's use to examine issued patents and records, supplies copies of records and other papers, and the like. Similar functions are performed with respect to the registration of trademarks.

The USPTO has no jurisdiction over questions of infringement and the enforcement of patents, nor over matters relating to the promotion or utilization of patents or inventions.

The work of examining applications for patents is divided among a number of examining technology centers (TCs), each TC having jurisdiction over certain assigned fields of technology. Each TC is headed by a group director and staffed by examiners. The examiners review applications for patents and determine whether patents can be granted. An appeal can be taken to the Patent Trial and Appeal Board regarding their decision to refuse to grant a patent, and review by the Director of the USPTO may be had on other matters by petition.

At present, the USPTO has over 12,000 employees, of whom about three quarters are examiners and others with technical and legal training. Patent applications are received at the rate of over five hundred thousand per year.

Some people confuse patents, copyrights, and trademarks. Although there may be some similarities among these kinds of intellectual property protection, they are different and serve different purposes.

What Is a Patent?

A patent for an invention is the grant of an intellectual property right to the inventor, issued by the Patent and Trademark Office. The term of a new patent is twenty years from the date on which the application for the patent was filed in the United States or, in special cases, from the date an earlier related application was filed, subject to the payment of maintenance fees. U.S. patent grants are effective only within the United States, its territories, and its possessions.

The right conferred by the patent grant is, in the language of the statute and of the grant itself, "the right to exclude others from making, using, offering for sale, or selling" the invention in the United States or "importing" the invention into the United States.

What is granted is not the right to make, use, offer for sale, sell or import, but the right to exclude others from making, using, offering for sale, selling or importing the invention. Once a patent is issued, the patentee must enforce the patent without aid of the USPTO.

What Is a Trademark or Service Mark?

A trademark is a word, name, symbol, or device which is used in trade with goods to indicate the source of the goods and to distinguish them from the goods of others. A service mark is the same as a trademark except that it identifies and distinguishes the source of a service rather than a product. The terms *trademark* and *mark* are commonly used to refer to both trademarks and service marks.

Trademark rights may be used to prevent others from using a confusingly similar mark, but not to prevent others from making the same goods or from selling the same goods or services under a clearly different mark. Trademarks which are used in interstate or foreign commerce may be registered with the USPTO. The registration procedure for trademarks and general information concerning trademarks is described in a pamphlet entitled "Basic Facts about Trademarks," available at the USPTO, and at www.uspto.gov/trademarks/basics/Basic_Facts_Trademarks.jsp.

What Is a Copyright?

A copyright is a form of protection provided to authors of "original works of authorship," including literary, dramatic, musical, artistic, and certain other intellectual works, both published and unpublished.

A copyright protects the form of expression rather than the subject matter of the writing. For example, a description of a machine could be copyrighted, but this would only prevent others from copying the description; it would not prevent others from writing a description of their own or from making and using the machine. Copyrights are registered by the Copyright Office of the Library of Congress. See more detail in the Resources section, or at www.copyright.gov.

What Can Be Patented?

The following is the USPTO's description of what can be patented. There are three types of patents.

- Utility patents may be granted to anyone who invents or discovers any new, useful, and unobvious process (primarily industrial or technical); machine; manufacture or composition of matter (generally chemical compounds, formulas, and the like); or any new, useful, and unobvious improvement thereof.
- Design patents may be granted to anyone who invents any new and unobvious original and ornamental design for an article of manufacture, such as a new auto body design. (Remember, a design patent may not always be valuable because a commercially similar design can easily be made without infringing the patent.)
- Plant patents may be granted to anyone who invents or discovers any distinct and new variety of plant, other than tuber-propagated, which is asexually reproduced.

In order for an invention to be patentable, it must be new as defined in the patent law, which provides that an invention cannot be patented if: "(a) the invention was known or used by others in this country, or patented or described in a printed publication in this or a foreign country, before the invention thereof by the applicant for patent," or "(b) the invention was patented or described in a printed publication or other public forum such as a public lecture, YouTube, or radio show, for example, in this or a foreign country or in public use or on sale in this country more than one year prior to the application for patent in the United States . . ."

If the invention has been described in a public forum anywhere, or has been in public use or on sale in this country more than one year before the date on which an application for patent is filed in this country, a patent cannot be obtained. In this connection it is immaterial when the invention was made, or whether the printed publication or public use was by the inventor himself/herself or by someone else. If the inventor describes the invention in a public forum or uses the invention publicly, or places it on sale, he/she must apply for a patent before one year has gone by; otherwise any right to a patent will be lost.

Even if the subject matter sought to be patented is not exactly shown by the prior art, and involves one or more differences over the most nearly similar thing already known, a patent may still be refused if the differences would be obvious. The subject matter sought to be patented must be sufficiently different from what has been used or described before it may be said to be nonobvious to a person having ordinary skill in the area of technology related to the invention. For example, the substitution of one material for another, or changes in size, are ordinarily not patentable.

Searches and Patent and Trademark Resource Centers

Many inventors attempt to make their own search of the prior patents and publications before applying for a patent. This may be done in the Public Search Facility of the USPTO, and in libraries located throughout the United States that have been designated as Patent and Trademark Resource Centers (PTRCs).

An inventor may also employ patent attorneys or agents to perform the preliminary search. This search may not be as complete as that made by the USPTO during the examination of an application and only serves, as its name indicates, a preliminary purpose. For this reason, the patent examiner may, and often does, reject claims in an application on the basis of prior patents or publications not found in the preliminary search. For a complete list of PTRCs, refer to the USPTO website at www.uspto.gov/products/library/ptdl/index.jsp.

The Scientific and Technical Information Center of the USPTO located at 1D58 Remsen Bldg. Lower Level, 400 Dulany Street, Alexandria, Virginia, has available for public use over 120,000 volumes of scientific and technical books in various languages, about 90,000 bound volumes of periodicals devoted to science and technology, the official journals of 77 foreign patent organizations, and over 40 million foreign patents on paper, microfilm, microfiche, and CD-ROM. The Scientific and Technical Information Center is open to the public from 8:00 A.M. to 5:00 P.M., Monday through Friday except federal holidays.

The Patent Search Room located at Madison East, First Floor, 600 Dulany Street, Alexandria, Virginia, is a place where the public may search and examine U.S. patents granted since 1790 using state-of-the-art computer workstations. Patents are arranged according to the U.S. Patent Classification System of over 450 classes and over 150,000 subclasses. By searching in these classified groupings of patents, it is possible to determine, before actually filing an application, whether an invention has been anticipated by a U.S. patent, and it is also possible to obtain the information contained in patents relating to any field of endeavor. The Patent Search Room (www.uspto.gov/products/library/) contains microfilm of all U.S. patents and a complete set of the *Official Gazette.*

The Patent and Trademark Resource Centers (PTRCs), located at libraries throughout the U.S., access current issues of U.S. and foreign patents and maintain collections of earlier issued patents and trademark information. The scope of these collections varies from library to library, ranging from patents of only recent years to all or most of the patents issued since 1790.

These patent collections are open to public use. Each of the PTRCs, in addition, offers the publications of the Office of Patent Classification (the *Manual of Classification, Index to the U.S. Patent Classification System,* and *Classification Definitions*), access to the CPC, and other patent documents and forms, and provides technical staff assistance to aid the public in gaining effective access to information contained in patents. The collections are organized in patent number sequence.

The Cassis CD-ROM system is available in all PTRCs. With various files, it permits the effective identification of appropriate classifications to search, provides numbers of patents assigned to a classification to facilitate finding the patents in a numerical file of patents, provides the current classification(s) of all patents, permits word searching on classification titles and on abstracts, and provides certain bibliographic information on more recently issued patents.

Due to variations in the scope of patent collections among the PTRCs and in their hours of service to the public, anyone contemplating the use of the patents at a particular library is advised to contact that library, in advance, about its collection, services, and hours, so as to avert possible inconvenience. See a list of PTRCs in Appendix B, or go to: www.uspto.gov/products/library/ptdl/locations/index.jsp.

Attorneys and Agents

The process of applying for and attaining a patent from the United States Patent and Trademark Office (USPTO or Office) is an undertaking that requires knowledge of patent law and rules; the Office's practices and procedures; and the scientific or technical matters involved in the particular invention.

Inventors may prepare their own applications, file them in the USPTO, and conduct the proceedings themselves; however, unless they are familiar with these matters or study them in detail, they may run into considerable difficulty. That is, there would be no assurance that the patent obtained would adequately protect the invention.

Most inventors employ the services of registered patent attorneys or patent agents. The Office has the power to make rules and regulations governing the conduct and recognition of patent attorneys and patent agents who practice before it. Persons who are not recognized by the USPTO for this practice are not permitted by law to represent inventors before the USPTO.

The Office registers both attorneys at law and persons who are not attorneys at law. The former persons are referred to as "patent attorneys" and the latter persons are referred to as "patent agents." Both patent attorneys and patent agents are permitted to prepare an application for a patent and conduct the prosecution in the USPTO. Patent agents, however, cannot conduct patent litigation in the courts or perform various services that the local jurisdiction considers practicing law. For example, a patent agent cannot draw up a contract relating to a patent if the state in which he/she resides considers that practicing law.

The USPTO maintains a register of patent attorneys and patent agents. To be admitted to this register, a person must comply with the regulations prescribed by the Office, which require proof that the person is of good moral character and good repute and that he/she has the legal, scientific, and technical qualifications necessary to offer a valuable service to patent applicants. Some of these qualifications must be demonstrated by passing an examination. Those admitted to the examination must have a college degree in engineering or physical science or the equivalent of such a degree.

The USPTO's directory of registered patent attorneys and patent agents is available at https://oedci.uspto.gov/OEDCI. The USPTO cannot recommend any particular attorney, agent, or firm or aid in the selection thereof (for instance, by stating that a certain party is "reliable" or "capable"). The classified section of telephone directories in most major cities also has a list of local patent attorneys under the heading "patent attorneys."

Some individuals and organizations that are not registered advertise their services in the fields of patent searching and invention marketing and development. Such individuals and organizations cannot represent inventors before the USPTO. They are not subject to USPTO discipline, but the USPTO does provide a public forum where complaints and comments about invention promoters and promotion firms are posted (www.uspto.gov/web/offices/com/iip/complaints.htm).

When employing a patent attorney or agent, the inventor executes a power of attorney that is filed in the USPTO and included in the application. Once a registered attorney or agent is appointed, the Office communicates directly with him/her rather than the inventor. However, the inventor is free to contact the USPTO concerning the status of his/her application and may remove the attorney or agent by revoking the power of attorney.

The USPTO has the power to disbar or suspend persons guilty of gross misconduct, but this can only be done after a full hearing with the presentation of clear and convincing evidence of the misconduct. The USPTO receives and, when appropriate, acts upon legitimate complaints against patent attorneys and agents.

The fees that patent attorneys and agents charge for their professional services are not regulated by the USPTO. Definitive evidence of overcharging may serve as a basis for USPTO action, but the Office rarely intervenes in disputes concerning fees.

Who May Apply for a Patent

According to the law, only the inventor may apply for a patent, with certain exceptions. If a person who is not the inventor should apply for a patent, the patent, if it were obtained, would be invalid. The person applying in such a case who falsely states that he/she is the inventor would also be subject to criminal penalties. If the inventor is dead, the application may be made by legal representatives, that is, the administrator or executor of the estate. If the inventor is insane, the application for patent may be made by a guardian. If an inventor refuses to apply for a patent or cannot be found, a joint inventor or a person having a proprietary interest in the invention may apply on behalf of the non-signing inventor.

If two or more persons make an invention jointly, they apply for a patent as joint inventors. A person who makes a financial contribution is not a joint inventor and cannot be joined in the application as an inventor. It is possible to correct an innocent mistake of erroneously omitting an inventor or erroneously naming a person as an inventor.

The Patent Application

With few exceptions, the patent application must be filed in the name of the inventor. Even the application for a patent on an invention by a company's researcher must be filed in the inventor's name. If there is more than one inventor, a joint application is made. The patent application can be assigned, however, to an individual or a corporation, and then the patent will be granted to the assignee, although filed in the inventor's name.

Often employment agreements require an employee to assign to the employer any invention relating to the employer's business. Even without such an agreement, the employer may have a "shop right" to use (free) an invention developed on the job by an employee.

Application for a patent is made to the Director of the U.S. Patent and Trademark Office, and includes:

- A written document that comprises a petition, a specification (descriptions and claims), and an oath
- A drawing in those cases in which a drawing is possible
- The current filing fee, generally less than $500

The construction of the invention, its operation, and its advantages should be accurately described. From the disclosure of the application, any person skilled in the field of the invention should be able to understand the intended construction and use of the invention. Commercial advantages, which would be attractive to a prospective manufacturer, need not be discussed.

The claims at the end of the specification point out the patentable new features of the invention. Drawings must be submitted according to rigid Patent and Trademark Office regulations.

Making Applications Special

Only under limited conditions may a petition be filed requesting that an application be given special treatment, that is, taken up for examination before its turn is reached. These requirements are of particular importance to small business owners who are eager to obtain a patent before starting a manufacturing program.

Here is a list of some of the conditions that may qualify your patent application to be treated as special. A petition to make an application special may be filed without a fee if the basis for the petition is (a) the applicant's age or health or (b) that the invention will materially enhance the quality of the environment, contribute to the development or conservation of energy resources, or contribute to countering terrorism.

Your attorney must file an affidavit to show that he or she has made a careful and thorough search of the prior art and believes that all the claims in the application are allowable. The attorney will also be expected to make sure that the sworn statement is properly filed. For more details about "special" application, see www.uspto.gov/web/offices/pac/mpep/s708.html.

Provisional Application for a Patent

Since June 8, 1995, the USPTO has offered inventors the option of filing a provisional application for patent, which was designed to provide a lower-cost first-patent filing in the United States and to give U.S. applicants parity with foreign applicants. Claims and oath or declaration are *not* required for a provisional application. Provisional application provides the means to establish an early effective filing date in a patent application and permits the term "Patent Pending" to be applied in connection with the invention. Provisional applications may not be filed for design inventions.

The filing date of a provisional application is the date on which a written description of the invention, drawings if necessary, and the name of the inventor(s) are received in the USPTO. The applicant would then have up to twelve months to file a non-provisional application for his or her patent as described above. The claimed subject matter in the later-filed non-provisional application is entitled to the benefit of the filing date of the provisional application if it has support in the provisional application.

Provisional applications are not examined on their merits. A provisional application will become abandoned by the operation of law twelve months from its filing date. The twelve-month pendency for a provisional application is not counted toward the twenty-year term of a patent granted on a subsequently filed non-provisional application, which relies on the filing date of the provisional application.

Examination of Applications and Proceedings

Applications, other than provisional applications, filed in the USPTO and accepted as complete applications are assigned for examination to the respective examining groups having charge of the areas of technology related to the invention. In the examining TC, applications are taken up for examination by the examiner to whom they have been assigned in the order in which they have been filed or in accordance with examining procedures established by the Director.

Applications will not be advanced out of turn for examination or for further action except as provided by the rules, or upon order of the Director to expedite the business of the Office, or upon a showing which, in the opinion of the Director, will justify advancing them. Making an application *special* is one such exception.

The examination of the application consists of a study of the application for compliance with the legal requirements and a search through U.S. patents, published patent applications, foreign patent documents, and available literature, to see if the claimed invention is new, useful, and nonobvious and if the application meets the requirements of the patent statute and rules of practice. The examiner reaches a decision in the light of the study and the result of the search.

If the examiner's decision is favorable, the patent is "allowed."

Office Action

The applicant is notified in writing of the examiner's decision by an "office action," which is normally mailed to the attorney or agent of record. The reasons for any adverse action or any objection or requirement are stated in the action, and such information or references that are given may be useful in aiding the applicant to judge the propriety of continuing the prosecution of his/her application.

If the claimed invention is not directed to patentable subject matter, the claims will be rejected. If the examiner finds that the claimed invention lacks novelty or differs only in an obvious manner from what is found in the prior art, the claims may also be rejected. It is not uncommon for some or all of the claims to be rejected on the first office action by the examiner; relatively few applications are allowed as filed.

Applicant's Reply

The applicant must request reconsideration in writing, and must distinctly and specifically point out the supposed errors in the examiner's office action. The applicant must reply to every ground of objection and rejection in the prior Office action (except that the applicant may request that objections or requirements as to form not necessary to further consideration of the claims be held in abeyance until allowable subject matter is indicated). The applicant's reply must appear throughout to be a bona fide attempt to advance the case to final action. The mere allegation that the examiner has erred will not be received as a proper reason for such reconsideration.

In amending an application in reply to a rejection, the applicant must clearly point out why he/she thinks the amended claims are patentable in view of the state of the art disclosed by the prior references cited or the objections made. He/she must also show how the claims as amended avoid such references or objections. After reply by the applicant, the application will be reconsidered, and the applicant will be notified as to the status of the claims, that is, whether the claims are rejected, or objected to, or whether the claims are allowed, in the same manner as after the first examination. The second Office action usually will be made final.

Final Rejection

On the second or later consideration, the rejection or other action may be made final. The applicant's reply is then limited to appeal in the case of rejection of any claim, and further amendment is restricted. Petition may be taken to the Director in the case of objections or requirements not involved in the rejection of any claim.

Interviews with examiners may be arranged, but an interview does not remove the necessity for reply to Office actions within the required time, and the action of the Office is based solely on the written record.

Restrictions

If two or more inventions are claimed in a single application, and are regarded by the Office to be of such a nature that a single patent should not be issued for both of them, the applicant will be required to limit the application to one of the inventions. The other invention may be made the subject of a separate application, which, if filed while the first application is still pending, will be entitled to the benefit of the filing date of the first application. A requirement to restrict the application to one invention may be made before further action by the examiner.

Appeals

If the examiner persists in the rejection of any of the claims in an application, or if the rejection has been made final, the applicant may appeal to the Patent Trial and Appeal Board (PTAB) in the USPTO. The PTAB consists of the Director of Patents and Trademarks, the Deputy Director, and the administrative patent judges, but normally each appeal is heard by only three members. An appeal fee is required, and the applicant must file a brief to support his/her position. An oral hearing will be held if requested upon payment of the specified fee.

There are other alternatives to appealing a final rejection of your application, including but not limited to Request for Continued Examination (RCE), a Continued Prosecution Application (CPA), and so forth.

If the decision of the PTAB is still adverse to the applicant, an appeal may be taken to the Court of Appeals for the Federal Circuit, or a civil action may be filed against the Director in the U.S. District Court for the District of Columbia. The Court of Appeals for the Federal Circuit will review the record made in the Office and may affirm or reverse the Office's action. In a civil action, the applicant may present testimony in the court, and the court will make a decision.

Allowance and Issue of Patent

If, on examination of the application or at a later stage during the reconsideration of the application, the patent application is found to be allowable, a notice of allowance will be sent to the applicant or to applicant's attorney or agent of record, if any, and a fee for issuing the patent is due within three months from the date of the notice. If timely payment of the issue fee is not made, the application will be regarded as abandoned. The issued patent is publicly published in the *Official Gazette*.

In case the publication of an invention by the granting of a patent would be detrimental to the national defense, the patent law gives the Director the power to withhold the grant of a patent and to order the invention be kept secret for such period of time as the national interest requires.

Publication of Patent Applications

Publication of patent applications is required for most plant and utility patent applications eighteen months after filing (or earlier claimed priority date), or under the Patent Cooperation Treaty. Following publication, the patent application is no longer held in confidence by the Patent Office, and any member of the public may request access to the entire file history of the application.

At the time of original patent filing, an applicant may request that the application *not* be published, but only if the invention has not been and may not be the subject of an application filed in a foreign country that requires publication.

As a result of publication, an applicant may assert provisional rights. These rights provide a patentee with the opportunity to obtain a reasonable royalty from a third party that infringes a published application claim, provided actual notice is given to the third party by applicant and the patent issues from the application with a substantially identical claim. Thus, damages for pre-patent grant infringement by another are now available.

The Nature of a Patent and Patent Rights

Since the patent does not grant the right to make, use, offer for sale, sell, or import the invention, the patentee's own right to do so is dependent upon the rights of others and whatever general laws might be applicable. A patentee, merely because he or she has received a patent for an invention, is not thereby authorized to make, use, offer for sale, sell, or import the invention if doing so would violate any law. An inventor of a new automobile who has obtained a patent thereon would not be entitled to use the patented automobile in violation of the laws of a state requiring a license, nor may a patentee sell an article, the sale of which may be forbidden by a law, merely because a patent has been obtained.

Neither may a patentee make, use, offer for sale, sell, or import his/her own invention if doing so would infringe the prior rights of others. A patentee may not violate the federal antitrust laws, such as by resale price agreements or entering into combination in restraints of trade, or the pure food and drug laws, by virtue of having a patent. Ordinarily there is nothing which prohibits a patentee from making, using, offering for sale, selling, or importing his/her own invention, unless he/she thereby infringes another's patent, which is still in force.

Maintenance Fees

All utility patents that issue from applications filed on and after December 12, 1980, are subject to the payment of maintenance fees, which must be paid to maintain the patent in force. These fees are due at 3½, 7½, and 11½ years from the date the patent is granted and can be paid without a surcharge during the window period, which is the six-month period preceding each due date, e.g., three years to three years and six months. (See fee schedule for a list of maintenance fees.)

Failure to pay the current maintenance fee on time may result in expiration of the patent. A six-month grace period is provided when the maintenance fee may be paid with a surcharge. The grace period is the six-month period immediately following the due date. The USPTO does not mail notices to patent owners that maintenance fees are due. If, however, the maintenance fee is not paid on time, efforts are made to remind the responsible party that the maintenance fee may be paid during the grace period with a surcharge.

Assignments and Licenses

A patent is personal property and may be sold to others or mortgaged; it may be bequeathed by a will, and it may pass to the heirs of a deceased patentee. The patent law provides for the transfer or sale of a patent, or of an application for a patent, by an instrument in writing. Such an instrument is referred to as an assignment and may transfer the entire interest in the patent. The statute also provides for the assignment of a part interest, that is, a half interest, a fourth interest, etc., in a patent.

An assignment, grant, or conveyance of any patent or application for a patent should be acknowledged before a notary public or officer authorized to administer oaths or perform notarial acts. The certificate of such acknowledgment constitutes prima facie evidence of the execution of the assignment, grant, or conveyance.

Joint Ownership

Patents may be owned jointly by two or more persons as in the case of a patent granted to joint inventors, or in the case of the assignment of a part interest in a patent. Any joint owner of a patent, no matter how small the part interest, may, without regard to the other owners, sell the interest or any part of it, or grant licenses to others, without regard to the other joint owner, unless the joint owners have made a contract governing their relation to each other.

The owner of a patent may also grant licenses to others. The drawing up of a license agreement (as well as assignments) is within the field of an attorney at law.

Such attorney should be familiar with patent matters as well. A few states have prescribed certain formalities to be observed in connection with the sale of patent rights.

Infringement of Patents

Suits for infringement of patents follow the rules of procedure of the federal courts. From the decision of the district court, there is an appeal to the Court of Appeals for the Federal Circuit. The Supreme Court may thereafter take a case by writ of certiorari. If the U.S. government infringes a patent, the patentee has a remedy for damages in the U.S. Court of Federal Claims. The government may use any patented invention without permission of the patentee, but the patentee is entitled to obtain compensation for the use by or for the government.

The Office has no jurisdiction over questions relating to infringement of patents. In examining applications for patent, no determination is made as to whether the invention sought to be patented infringes any prior patent. An improvement invention may be patentable, but it might infringe a prior unexpired patent for the invention improved upon, if there is one.

Patent Marking and Patent Pending

A patentee who makes or sells patented articles, or a person who does so for or under the patentee is required to mark the articles with the word "Patent" and the number of the patent. The penalty for failure to mark is that the patentee may not recover damages from an infringer unless the infringer was duly notified of the infringement and continued to infringe after the notice.

The marking of an article as patented when it is not in fact patented is against the law and subjects the offender to a penalty. Some persons mark articles sold with the terms "Patent Applied For" or "Patent Pending." These phrases have no legal effect, but only give information that an application for patent has been filed in the USPTO. The protection afforded by a patent does not start until the actual grant of the patent. False use of these phrases or their equivalent is prohibited.

Design Patents

The patent laws provide for the granting of design patents to any person who has invented any new and nonobvious ornamental design for an article of manufacture. The design patent protects only the appearance of an article, but not its structural or functional features. The proceedings relating to granting of design patents are the same as those relating to other patents with a few differences. A design patent has a term of fourteen years from grant, and no fee is necessary to maintain a design patent in force. The drawing of the design patent conforms to the same rules as other drawings, but no reference characters are allowed, and the drawing should clearly depict the appearance, since the drawing defines the scope of patent protection. The specification of a design application is short and ordinarily follows a set form. Only one claim is permitted, following a set form.

Plant Patents

The law also provides for the granting of a patent to anyone who has invented or discovered and asexually reproduced any distinct and new variety of plant, including cultivated sports, mutants, hybrids, and newly found seedlings, other than a tuber-propagated plant or a plant found in an uncultivated state.

Asexually propagated plants are those that are reproduced by means other than from seeds, such as by the rooting of cuttings, by layering, budding, grafting, inarching, etc. A plant patent is granted on the entire plant. It therefore follows that only one claim is necessary and only one is permitted. All inquiries relating to plant patents and pending plant patent applications should be directed to the USPTO and not to the Department of Agriculture.

The Plant Variety Protection Act (Public Law 91577), approved December 24, 1970, provides for a system of protection for sexually reproduced varieties, for which protection was not previously provided, under the administration of a Plant Variety Protection Office within the Department of Agriculture. Requests for information regarding the protection of sexually reproduced varieties should be addressed to Commissioner, Plant Variety Protection Office, Agricultural Marketing Service, National Agricultural Library Bldg., Room 0, 10301 Baltimore Boulevard, Beltsville, Maryland 20705-2351.

Treaties and Foreign Patents

Since the rights granted by a U.S. patent extend only throughout the territory of the United States and have no effect in a foreign country, an inventor who wishes patent protection in other countries must apply for a patent in each of the other countries or in regional patent offices. Almost every country has its own patent law, and a person desiring a patent in a particular country must make an application for patent in that country, in accordance with the requirements of that country.

The laws of many countries differ in various respects from the patent law of the United States. In most foreign countries, publication of the invention before the date of the application will bar the right to a patent. In most foreign countries, maintenance fees are required. Most foreign countries require that the patented invention must be manufactured in that country after a certain period, usually three years. If there is no manufacture within this period, the patent may be void in some countries, although in most countries the patent may be subject to the grant of compulsory licenses to any person who may apply for a license.

There is a treaty relating to patents which is adhered to by 175 countries, including the United States, and is known as the Paris Convention for the Protection of Industrial Property. It provides that each country guarantees to the citizens of the other countries the same rights in patent and trademark matters that it gives to its own citizens. The treaty also provides for the right of priority in the case of patents, trademarks, and industrial designs (design patents). This right means that, on the basis of a regular first application filed in one of the member countries, the applicant may, within a certain period of time, apply for protection in all the other member countries. These later applications will then be regarded as if they had been filed on the same day as the first application. Thus, these later applicants will have priority over applications for the same invention which may have been filed during the same period of time by other persons. Moreover, these later applications, being based on the first application, will not be invalidated by any acts accomplished in the interval, such as, for example, publication or exploitation of the invention, the sale of copies of the design, or use of the trademark. The period of time mentioned above, within which the subsequent applications may be filed in the other countries, is twelve months in the case of first applications for patent and six months in the case of industrial designs and trademarks.

Another treaty, known as the Patent Cooperation Treaty (PCT), is presently adhered to by over 148 countries, including the United States. The treaty facilitates the filing of applications for patent on the same invention in member countries by providing, among other things, for centralized filing procedures and a standardized application format.

The timely filing of an international application affords applicants an international filing date in each country, which is designated in the international application and provides a search of the invention and a later time period within which the national applications for patent must be filed.

Under U.S. law it is necessary, in the case of inventions made in the United States, to obtain a license from the Director of the USPTO before applying for a patent in a foreign country. Such a license is required if the foreign application is to be filed before an application is filed in the United States or before the expiration of six months from the filing of an application in the United States. The filing of an application for patent constitutes the request for a license, and the granting or denial of such request is indicated in the filing receipt mailed to each applicant. After six months from the U.S. filing, a license is not required unless the invention has been ordered to be kept secret. If the invention has been ordered to be kept secret, consent to file abroad must be obtained from the Director of the USPTO during the period the order of secrecy is in effect.

Foreign Applicants for U.S. Patents

The patent laws of the United States make no discrimination with respect to the citizenship of the inventor. Any inventor, regardless of his/her citizenship, may apply for a patent on the same basis as a U.S. citizen.

A foreign applicant may be represented by any patent attorney or agent who is registered to practice before the U.S. Patent and Trademark Office.

This material was excerpted from *General Information Concerning Patents* produced by the USPTO, with minor revision by the author. It is periodically updated to reflect changes in the laws, procedures, and contact information. For updates and further information, contact:

Independent Inventor Assistance

USPTO Independent Inventor Program
Mail Stop 24
P.O. Box 1450
Alexandria, VA 22313-1450
Toll free: 866-767-8877
Phone: 571-272-8033
usptoinfo@uspto.gov
www.uspto.gov/inventors/

General Inquiries

Toll free: 800-786-9199
Phone: 571-272-1000
TTY: 571-272-9950
usptoinfo@uspto.gov
www.uspto.gov

The principal locations of the USPTO are in Alexandria, Virginia.

APPENDIX B

Patent and Trademark Resource Centers

(Formerly known as U.S. Patent and Trademark Depository Libraries (USPTDLs))
For updates about PTRCs go to: www.uspto.gov/products/library/ptdl/locations/

ALABAMA

Auburn
Ralph Brown Draughon Library
Auburn University
334-844-1737

Birmingham
Birmingham Public Library
205-226-3620

ALASKA

Fairbanks
Mather Library
Geophysical Institute, University of Alaska
907-474-2636

ARIZONA

Phoenix
State Library of Arizona
602-926-3870

ARKANSAS

Little Rock
Arkansas State Library
501-682-2053

CALIFORNIA

Los Angeles
Los Angeles Public Library
213-228-7220

Riverside
Orbach Science Library
University of California, Riverside
951-827-3316

San Diego
San Diego Public Library
619-236-5800

San Francisco
San Francisco Public Library
415-557-4400

Sunnyvale
Sunnyvale Public Library
408-730-7300

COLORADO

Denver
Denver Public Library
720-865-1711

CONNECTICUT

Fairfield
Ryan-Matura Library
Sacred Heart University
203-371-7726

DELAWARE

Newark
University of Delaware Library
302-831-2965

DISTRICT OF COLUMBIA

Washington
Founders Library
Howard University
202-806-7252

FLORIDA

Fort Lauderdale
Broward County Main Library
954-357-7444

Miami
Miami-Dade Public Library
305-375-2665

ORLANDO
University of Central Florida Library
407-823-2562

GEORGIA

ATLANTA
Georgia Tech Library
404-385-7185

HAWAII

HONOLULU
Hawaii State Library
808-586-3477

ILLINOIS

CHICAGO
Chicago Public Library
312-747-4450

MACOMB
Leslie S. Malpass Library
Western Illinois University
309-298-2722

INDIANA

INDIANAPOLIS
Indianapolis-Marion County Public
317-275-4100

WEST LAFAYETTE
Siegesmund Engineering Library
Purdue University
765-494-2872

IOWA

DAVENPORT
Davenport Public Library
563-326-7832

KANSAS

WICHITA
Ablah Library
Wichita State University
800-572-8368

KENTUCKY

HIGHLAND HEIGHTS
W. Frank Steely Library
Northern Kentucky University
859-572-5457

LOUISVILLE
Louisville Free Public Library
502-574-1611

LOUISIANA

BATON ROUGE
Troy H. Middleton Library
Louisiana State University
225-578-8875

MAINE

ORONO
Raymond H. Fogler Library
University of Maine
207-581-1678

MARYLAND

BALTIMORE
University of Baltimore Law Library
410-837-4554

COLLEGE PARK
Engineering & Physical Sciences Library
University of Maryland
301-405-9157

MASSACHUSETTS

AMHERST
Science and Engineering Library
University of Massachusetts at Amherst
413-545-1370

BOSTON
Boston Public Library
617-536-5400, Ext. 2226

MICHIGAN

ANN ARBOR
Art, Architecture & Engineering Library
University of Michigan
734-647-5735

BIG RAPIDS
Ferris Library of Information, Technology, and Education
Ferris State University
231-591-3602

Detroit
Detroit Public Library
313-481-1391

Houghton
Van Pelt and Opie Library
Michigan Technological University
906-487-2500

MINNESOTA

Minneapolis
Hennepin County Library
Minneapolis Central
612-543-8000

MISSISSIPPI

Jackson
Mississippi Library Commission
601-432-4111

MISSOURI

Kansas City
Linda Hall Library
816-363-4600, Ext. 724

St. Louis
St. Louis Public Library
314-539-0390

MONTANA

Butte
Montana Tech Library
University of Montana
406-496-4281

NEBRASKA

Lincoln
Engineering Library
University of Nebraska, Lincoln
402-472-3411

NEVADA

Reno
University Library
University of Nevada, Reno
775-682-5593

NEW HAMPSHIRE

Concord
University of New Hampshire
School of Law Library
603-513-5130

NEW JERSEY

Newark
Newark Public Library
973-733-7779

Piscataway
Library of Science and Medicine
Rutgers University
848-445-2895

NEW YORK

Albany
New York State Library
518-474-5355

Buffalo
Buffalo and Erie County
Public Library
716-858-8900

New York
Science, Industry, & Business Library
New York Public Library
212-592-7000

Rochester
Central Library of Rochester
and Monroe
585-428-8110

Smithtown
Smithtown Main Library
631-265-2072

NORTH CAROLINA

Charlotte
J. Murrey Atkins Library
University of North Carolina, Charlotte
704-687-0494

Raleigh
D. H. Hill Library
North Carolina State University
919-515-6602

NORTH DAKOTA

GRAND FORKS

Chester Fritz Library
University of North Dakota
701-777-4888

OHIO

AKRON

Akron-Summit County Public Library
330-643-9075

CINCINNATI

Cincinnati & Hamilton Public Library
513-369-6932

CLEVELAND

Cleveland Public Library
216-623-2870

DAYTON

Wright State University
937-775-3521

TOLEDO

Toledo/Lucas County Public Library
419-259-5209

OKLAHOMA

STILLWATER

Edmond Low Library
Oklahoma State University
405-744-6546

PENNSYLVANIA

PHILADELPHIA

The Free Library of Philadelphia
215-686-5394

PITTSBURGH

The Carnegie Library of Pittsburgh
412-622-3138

UNIVERSITY PARK

PAMS Library
Pennsylvania State University
814-865-7617

PUERTO RICO

BAYAMÓN

Learning Resource Center, Bayamón Campus
University of Puerto Rico
787-993-0000, Ext. 3222

MAYAGÜEZ

General Library, Mayagüez Campus
University of Puerto Rico
787-832-4040, Ext. 2307

RHODE ISLAND

PROVIDENCE

Providence Public Library
401-455-8027

SOUTH CAROLINA

CLEMSON

R. M. Cooper Library
Clemson University
864-656-3024

SOUTH DAKOTA

RAPID CITY

Devereaux Library
South Dakota School of Mines and Technology
605-394-1275

TENNESSEE

NASHVILLE

Stevenson Science & Engineering Library
Vanderbilt University
615-322-2717

TEXAS

AUSTIN

McKinney Engineering Library
University of Texas at Austin
512-495-4511

COLLEGE STATION

West Campus Library
Texas A&M University
979-845-2111

Dallas
Dallas Public Library
214-670-1468

Houston
Fondren Library
Rice University
713-348-5483

Lubbock
Texas Tech University Library
806-742-2282

San Antonio
San Antonio Public Library
210-207-2500

UTAH

Salt Lake City
Marriott Library
University of Utah
801-581-8394

VERMONT

Burlington
Bailey/Howe Library
University of Vermont
802-656-2542

WASHINGTON

Seattle
Engineering Library
University of Washington
206-543-0740

WEST VIRGINIA

Morgantown
Evansdale Library
West Virginia University
304-293-4695

WISCONSIN

Madison
Wendt Commons Library,
University of Wisconsin-Madison
608-265-9802

Milwaukee
Milwaukee Public Library
414-286-3051

WYOMING

Cheyenne
Wyoming State Library
307-777-7281

APPENDIX C

INVENTION EVALUATION CRITERIA

The PIES-VIII evaluation format created by Gerald G. Udell, Ph.D., at The Innovation Institute, is the product of more than twenty years' experience and research. The forty-one criteria have been selected and modified over the years to present a profile of the risks involved in commercializing an invention. This profile is intended to serve as an aid in decision making, suggest strategies for further development, and help inventors better understand the complexities of the project innovation process.

SOCIAL CRITERIA

1. *Legality:* In terms of applicable laws (particularly product liability), regulations, and product standards
2. *Safety:* Considering potential hazards and side effects
3. *Environmental Impact:* In terms of pollution, litter, misuse of natural resources, etc.
4. *Societal Impact:* In terms of the impact (benefit) upon the general welfare of society

BUSINESS RISK CRITERIA

5. *Functional Feasibility:* In terms of intended functions, will it actually do what it is intended to do?
6. *Production Feasibility:* With regard to technical process or equipment required for production
7. *Stage of Development:* Based on available information
8. *Investment Costs:* The amount of capital needed for commercialization
9. *Payback Period:* Given adequate financing, the expected payback period (time required to recover initial investment after full market introduction)
10. *Profitability:* Defined as the extent to which anticipated revenues will cover the relevant costs (direct, indirect, and capital)
11. *Marketing Research:* The marketing research required for commercialization
12. *Research and Development:* The research and development required to reach the production-ready stage

DEMAND ANALYSIS CRITERIA

13. *Potential Market:* The total market for products of this type
14. *Potential Sales:* Given alternative and/or competitive products and the current market environment
15. *Trend of Demand:* The market demand for products of this type
16. *Stability of Demand:* The fluctuation in demand
17. *Product Life Cycle:* The product life cycle
18. *Product Line Potential:* The potential for additional products, multiple styles, qualities, price ranges, etc.

MARKET ACCEPTANCE CRITERIA

19. *Compatibility:* Compatibility with existing attitudes and methods of use
20. *Learning:* The amount of learning required for correct use
21. *Need:* The level of need filled or utility provided by this innovation
22. *Dependence:* The degree to which the sale or use of this product is dependent upon other products, processes, or systems
23. *Visibility:* The potential buyers, product advantages, and benefits
24. *Promotion:* Considering the market environment, the costs and effort required to promote the advantages, features, and benefits
25. *Distribution:* The cost and difficulty of establishing distribution channels
26. *Service:* The cost and difficulty associated with providing product service

COMPETITIVE CRITERIA

27. *Appearance:* Relative to relevant prior art, competition, and/or substitutes, how the appearance might be perceived by potential buyers
28. *Function:* Relative to relevant prior art, competing and/or substitute products, and services or processes, how the function performed might be perceived by potential buyers
29. *Durability:* Relative to relevant prior art, competition, and/or substitutes, how the durability of this product might be perceived by potential buyers
30. *Price:* Relative to relevant prior art, competition, and/or substitute products
31. *Existing Competition:* Existing competition, including both direct and indirect competition, and prior art for this innovation
32. *New Competition:* Competition from new entrant or competitive reaction
33. *Protection:* Considering patents, other prior art, copyrights, technical difficulty, secrecy, or other bars to competitive reaction, the prospects for protections

EXPERIENCE AND STRATEGY CRITERIA

34. *Technology Transfer:* The potential for technology (licensing) of this invention
35. *New Venture:* The potential for using this invention to start a new business venture
36. *Marketing Experience:* Commercialization of this idea/invention will require . . .
37. *Technical Experience:* Commercialization of this idea/invention will require . . .
38. *Financial Experience:* Commercialization of this ideal invention will require . . .
39. *Management/Production Experience:* Commercialization of this idea/invention will require . . .
40. *Initial Distribution Strategy:* The optimal channels for introducing this idea/invention into the marketplace
41. *Overall Market Attractiveness:* Considering competition, difficulty of establishing channels of distribution, cost of reaching target market, financial, technical and other barriers, and potential risk versus potential profit, the associated risk-profit ratio of this opportunity

Machines for printing paper hangings, cloth, books by Peter Force (U.S. Patent No. 3573X), Washington, D.C., 1822.

APPENDIX D

Risk/Reward Ratio Chart

There are several important things to bear in mind when considering this chart:

- Regardless of the score you get from this chart, your actual results may be entirely the opposite. The score may suggest extreme high risk and low chance for reward, and yet your project could be successful. On the other hand, it may show great potential and little risk, and your project could still fail. Take tests like this with a grain of salt. This chart does help you to understand the elements that go into invention projects, and which elements of risk you may need to work on.
- Things change. And they can change fast. You may be at more risk now because you aren't sure about patentability; however, with patent claims allowed, this could all change to your favor. Likewise, the knowledge gained from one call to a key executive could diminish your project from fantastic potential to useless.
- Higher scores are favorable because they equate to lower risk or greater chance for gain.
- If your exact circumstance doesn't appear on the chart, pick something close.

PATENT POSITION

(Select only one item to score in each category)

U.S. and key foreign patents issued	10
All possible competitors signed a Confidential Disclosure Agreement (CDA)	10
Key competitors signed CDA	9
U.S. patent issued and PCT filed	9
U.S. patent with broad claims	8
U.S. patent in uncrowded field	8
U.S. patent with narrow claims	7
U.S. patent in crowded field	7
Patent pending with claims allowed	7
Patent pending, no allowance	5
Favorable patent office search	4
Provisional pending or database search only	3
Desirable trademark and/or name	3
Design patent	2
Unfavorable search	1
Not patentable (except technical)	1
Nothing done	0

Score: _____

PROOF OF CONCEPT

Any necessary prototypes, tests, or accreditation completed	7
Uses standard engineering principal and readily available materials	6
Requires simple working prototype	5
Requires expensive working prototype or scientific evidence	3
Requires engineering tests or accreditations	2
Requires extensive tests or medical research	1

Score: _____

AVAILABLE RESOURCES FOR PATENTING, MARKETING, AND DEVELOPMENT

Funds in excess of $10,000 are readily available	7
Funds in excess of $5,000 are readily available	6
Funds in excess of $3,000 are readily available	5
Funds in excess of $1,500 are readily available	4
You can afford a patent search only ($200–$500)	2
No funds are available	1

Score: _____

MARKET

Consumable product	10
High-volume industrial product or process	9
Important medical or required regulatory	8
Mass appeal, i.e., Kmart, Wal-Mart	8
High-appeal niche, i.e., Home Depot, Best Buy	7
Medium niche, i.e., plumbing or electrical	5
Small niche, i.e., catalogs or specialties	4
Industrial medium niche	3
Industrial small niche	2
Score:	_____

RESISTANCE

Easily recognizable to consumers and the trade	8
Requires explanation, but is an easy concept	6
Requires explanation, is radically different	3
Requires scientific or engineering explanation	1
Score:	_____
Your Total Score:	_____

SCORE EVALUATION

Most desirable Risk/Reward score:
A score of 6 or higher in all categories and a total greater than 35.

Medium score:
A score of 4 or higher in all categories and total of 20 to 35.

Low score:
All scores under 4 or a total under 20.

Conclusion

If your score is less than 20, then the risks outweigh the chance for reward. Unless you can take measures to reduce your risks, there is an excellent chance that anything invested in your project will be a loss.

Even with a score in the 20 to 35 range, the risk/reward ratio is still vulnerable. Yet anything in this range is fair game for commercialization, possibly without much additional investment. For example, you may have a patented niche item that is easily understood and requires no prototype. In this event, being successful could be as simple as submitting your patent to the right company and accepting its offer. Granted things don't happen this smoothly very often, but it does happen.

Although a score of 35 or higher is definitely favorable, you're not necessarily out of the woods. You might have a simple and easily recognizable consumable product with ample financial resources and a broad patent. However, if too few people want it, or if the price is too high, you're up the creek, with no paddle I might add. Your score of 42 might as well be 0.

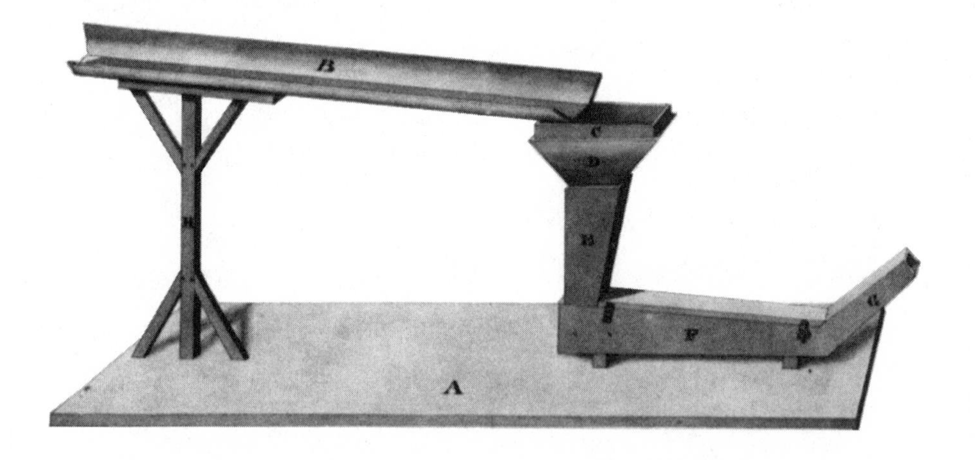

Washing, clearing, and separating gold dust by Richard Lee (U.S. Patent No. 5562X), Erwinsville, North Carolina, 1829.

APPENDIX E

SIMPLIFIED STEPS

A Quick-Reference Flow Chart of the Invention Commercialization Process

This chart is a quick-reference guide that gives an overview of the typical flow of an invention, from idea to commercialization (by using licensing as the commercialization method). Refer to specific sections within the book (e.g., Patent Searching, Market Research, Submitting Your Invention, Forming a Strategy, Negotiation, and so on) for in-depth instructions about how to effectively accomplish each step.

Non-Patented Idea or Invention	Patented Invention

DOCUMENT YOUR CONCEPTION DATE

- At a minimum, hand sketch and describe your invention, and sign and date each page.
- Have at least one or two people (not immediate family) write as follows: "I have read and understand the above." Have them sign and date their statement. Save this in a safe place.

PATENT AND PRIOR ART SEARCH

• Internet Search: http://patft.uspto.gov or www.google.com/patents	Free
• U.S. Patent and Trademark Depository Library	Free
• Patent Agent	$300 to $500
• Patent Attorney	$500 to $1,200

SEARCH RESULTS

- You find an invention that is identical or very similar to yours.
- You discover no identical invention, but yours is likely not patentable, or an obtainable patent would be very weak.
- You find nothing similar, and yours is very patentable, with a utility patent.

BACK TO THE DRAWING BOARD

- Modify or improve your design and start again from the top.
- In the event your invention is strong and unique enough to command the interest of a company that will enter into a confidential relationship and pay you for an unpatented idea, continue with the steps in the column to the right. Chances here are slimmer and licensing choices fewer, but not impossible.

ASSESS MARKETABILITY AND DETERMINE YOUR MARKET NICHE

- Do market research in the field.
- Interview industry experts.
- Check specialty catalogs and search the Internet.

IDENTIFY AND QUALIFY POTENTIAL LICENSEES

- Order catalogs, annual reports, SEC 10-K's, and product flyers.
- Redetermine that your invention is still unique.

ENTER INTO CONFIDENTIAL DISCLOSURE AGREEMENTS (CDAS)

- With potential licensees.
- Consider filing a patent application only if you receive positive responses in the following steps.
- Use DIMWIT.com to make a non-confidential disclosure.

FILE A PATENT

Cost: Full Utility Patent $1,500 to $4,000+
Provisional Patent $300 to $2,500

- Do this only if major multinational corporations are the only ones that qualify as potential licensees, then file a provisional or full patent application with the USPTO. Otherwise, you may avoid filing for a patent at this stage providing you have no impending time bars (consult a patent agent or attorney) and if you can keep all

ENTER INTO CONFIDENTIAL DISCLOSURE AGREEMENTS (CDAS)

- Only do this if confidential aspects or trade secrets are as important as patentable subject matter.
- Or enter into non-disclosure or non-confidential agreements as required by the company, and initially reveal only limited and non-confidential information.
- Or use DIMWIT.com to make a non-confidential disclosure.

SUBMIT YOUR IDEA

- With small- to medium-sized companies, contact a key decision maker first, and determine their interest in your basic concept.
- Submit to one potential licensee at a time, starting with your top candidate. Follow-up within thirty days.
- If you are rejected, always phone the reviewer and ask the reason for rejection, ask whether your invention has commercial possibilities, and get a referral to another potential licensee.
- If no one is interested, determine whether it is because of your invention or because you have targeted the wrong companies.
- Retarget potential licensees as necessary.

LICENSE STRATEGY REVIEW

- Determine the potential licensees' performance standards and territory.
- Choose a licensee.

NEGOTIATE AN EXCLUSIVE LICENSE

Cost: $500 to $3,000

- This has possible time limits on exclusivity and royalty reduction in the event competition arises.
- Refer to Chapter 6, and consult or retain a qualified patent attorney with licensing and preferably litigation experience.

- If the company is interested, consider negotiating a CDA to cover know-how and trade secrets to allow more in-depth communication.
- Otherwise, rely on your patent and/or potentially patentable aspects at this stage, but don't reveal which aspects are patentable. Don't reveal the proposed claims of your patent application or the application number.

DEVELOP YOUR STRATEGY

- Determine market segments, territory, exclusivity, and special requirements, and choose the best licensee candidate(s).

NEGOTIATE A LICENSE

Cost: $500 to $3,000

- Refer to Chapter 6 and consult or retain a qualified patent attorney with licensing and preferably litigation experience.

MONITOR THE MARKET AND QUALITY CONTROL

- Assist your licensee by being an extra set of eyes and ears in your industry.
- Listen to feedback and take pride in sharing that you are the inventor!
- Watch for follow-up opportunities, opportunities for improvements, and warning signs.

APPENDIX F

Disclosure Agreement

Printing machine by C. F. Vorheis (U.S. Patent No. 8139X), 1834.

Disclosure Agreement

Beginning _____________ , ____("INVENTOR") and/or his/her agents or assigns will disclose know-how, patented and/or unpatented inventions, designs, trade secrets, etc. ("INFORMATION"), which INVENTOR considers to be a valuable commercial asset, relating to:

The purpose of the disclosure is to allow confidential disclosure and communications between ____("COMPANY") and INVENTOR to discuss the development, marketing, and other relevant issues with respect to said INFORMATION. This is not an offer to sell or license.

1) It is understood that no obligation, express or implied, is assumed by the COMPANY unless and until a formal written contract has been entered into, and the obligation of the COMPANY shall be only such as is expressed in the formal written contract. INFORMATION disclosed to COMPANY shall remain property of INVENTOR.

2) COMPANY agrees that the disclosed INFORMATION will be held in strict confidence, and INFORMATION will not be disclosed to any other persons outside COMPANY, or used by COMPANY, without prior written consent from INVENTOR. INFORMATION released within COMPANY shall be on a need-to-know basis.

3) COMPANY shall not be bound to secrecy when COMPANY can document that said INFORMATION was: 1) previously developed by COMPANY, its divisions, subsidiaries, or affiliates, or, 2) disclosed to COMPANY by a third party completely independent of this or prior disclosures by INVENTOR, or, 3) if said INFORMATION is found in the public domain.

These terms are agreed to by:

____

Signature   Date

____

Name   Title

____

Company

RESOURCES

Sometimes one of the greatest problems for an inventor is knowing whom to trust and where to find reliable information. Historically I've found the following resources to be good sources of solid information that's particularly useful to independent inventors. This is by no means a complete list; however, it does present a comprehensive sample of the kinds of services available to inventors. The multitudes of other useful organizations can largely be accessed through the resources I've included here. I've annotated the organizations with which I have personal experience or relationships.

Note: When this symbol ✎ follows the title of a resource, it denotes services or information that may be of special interest to students.

This resources section is broken down into the following larger categories:

- Government Contacts
- Government Programs
- Patent Searching and Patenting
- Research Resources
- Small Business Resources
- Inventor Resources
- Crowdsourcing, Crowdfunding, and Open Innovation

Government Contacts

U.S. PATENT AND TRADEMARK OFFICE (USPTO)

Offers free access to patents, trademarks, lists of registered patent practitioners, and other searchable databases through its main website.

U.S. Patent and Trademark Office
P.O. Box 1450
Alexandria, VA 22313-1450

Toll free: 800-PTO-9199
Phone: 571-272-1000
uspto.info@uspto.gov
www.uspto.gov

U.S. PATENT AND TRADEMARK OFFICE (USPTO)—INVENTORS RESOURCES 

The USPTO offers special outreach programs to independent inventors. It hosts an annual inventors' conference, has a resource-rich website complete with a kids' page at www.uspto.gov/kids.

USPTO—Inventors Assistance Center
Mail Stop 24
P.O. Box 1450
Alexandria, VA 22313-1450

Toll Free: 800-PTO-9199
Phone: 571-272-1000
uspto.info@uspto.gov
www.uspto.gov/inventors

U.S. COPYRIGHT OFFICE

The copyright office located at the Library of Congress houses a tremendous amount of material from books to art. A copy of every piece of registered work is available here for public inspection. For information about registering your copyright, contact:

U.S. Copyright Office
Library of Congress
101 Independence Avenue SE
Washington, DC 20559-6000

Phone: 202-707-3000
Forms and publications hotline:
877-476-0778
www.copyright.gov

Government Programs

There are hundreds of programs available through state and federal governments. Key word searches on the Internet are an excellent way to identify sources of information from the U.S. government as well as the private sector. Since government programs and the funding for them are changing so rapidly, we will not dwell on identifying all of the specific government programs. The important thing is for you to know that many programs exist and to know how to reach them.

Each year the U.S. government, and some state governments, gives millions of dollars to independent inventors for the development of inventions. Although the U.S. Department of Energy (USDOE or DOE) has been most active in helping independent inventors, most areas of the government allocate money for inventors. If an inventor is inventing in an area that the U.S. or state government believes should be encouraged, they will solicit research proposals relating to those topics. Visit www.energy.gov/phonebook to find any phone number at USDOE.

USDOE (U.S. DEPARTMENT OF ENERGY)—GOLDEN FIELD OFFICE

The Golden Field Office works to bring to the world energy efficiency and renewable energy technologies, such as wind and solar power. As the primary field agent for the U.S. Department of Energy's Office of Energy Efficiency and Renewable Energy

(EERE), Golden builds partnerships to develop, commercialize, and encourage the use of those technologies and in doing so works closely with DOE's National Renewable Energy Laboratory (NREL), other national laboratories, the private sector, state and local governments, and many other stakeholders across the nation.

Golden is also responsible for administering the contract for the management and operation of NREL, the nation's primary laboratory for research and development into energy efficiency and renewable energy technologies.

U.S. Department of Energy
Golden Field Office
15013 Denver West Parkway
Golden, CO 80401

Phone: 720-356-1800
www.energy.gov/eere/about-us/business-operations/golden-field-office

IDEA (INNOVATIONS DESERVING EXPLORATORY ANALYSIS) PROGRAMS

IDEA programs encourage investigation of innovative concepts with potential for technological breakthroughs in transportation. Research contracts, averaging around $100,000, are awarded to individuals, companies, organizations, and universities that IDEA oversight committees feel will demonstrate new technologies or methods for improving the safety and efficiency of surface transportation in the United States in the following categories: National Cooperative Highway Research Program (NCHRP), highway, transit, and rail safety.

Transportation Research Board
The National Academies
500 Fifth Street NW
Washington, DC 20001

Phone: 202-334-3310
www.trb.org/IDEAProgram/IDEAProgram.aspx

THE CATALOG OF FEDERAL DOMESTIC ASSISTANCE (CFDA)

CFDA is a government-wide compendium of federal programs, projects, services, and activities that provides assistance or benefits to the American public. The PDF file (more than two thousand pages long) contains information about hundreds of project grants and financial and nonfinancial assistance programs administered by departments and establishments of the federal government. Find it at www.cfda.gov.

GRANTS.GOV

Grants.gov (www.grants.gov) allows organizations to electronically find and apply for competitive grant opportunities from all federal grant-making agencies. Grants.gov is THE single access point for over nine hundred grant programs offered by the twenty-six federal grant-making agencies.

MORE INFORMATION ABOUT GOVERNMENT GRANTS

One of the key things that the government does not want to do is compete with, or be construed as competing with, private enterprise. The last thing government employees need is a complaint to a congressperson by someone in private enterprise that they are being competed against by the U.S. government. Therefore, funds offered by the government are generally earmarked for research and development activities to advance and prove a concept so it would then be ready to turn over to private enterprise for commercialization.

This is an important distinction to realize. Many times inventors expect the government to help them with commercialization. Although this is not the role of government, the benefit to inventors does come indirectly since the research and development of an inventor's project may be the missing link necessary to prove a concept to a corporation that otherwise would not be willing to commercialize the project. Most inventors who have not had any experience using government resources view the government as a huge bureaucratic entity with no personality and foreign procedures. Although the government itself is very large, the specific programs that deal with inventors may only have a handful of employees. It is possible to get very personalized service. If there is something I don't know procedurally, I simply contact the right person and ask. Government employees in the area of invention are generally more than willing to point you in the right direction so you can proceed with your project.

It should be noted that programs such as the ones mentioned above are grants and not loans. You do not have to return grant money, as long as you abide by their rules and successfully complete the project. The federal government does not typically demand that you sign your rights over to the government; instead it generally reserves the rights to your invention for government use. You then have complete freedom to commercialize your invention in the private sector on your own. It didn't used to be this way. The U.S. government has loosened up over the years, and there is now a big push to move all types of government technologies into the private sector.

APPLYING FOR A GRANT

I once applied for a federal grant and was rejected because my grant proposal read more like a business plan. The second time around I got smarter and hired a person skilled in writing Small Business Innovation Research (SBIR) program grant proposals (see Small Business Resources in this section). The grant was awarded. The lesson here is that even though you may become chums with the government official who oversees your grant area, your written proposal still must be able to jump through the bureaucratic hoops. The people who review your grant proposal will likely be complete strangers to you. The government does this to help depersonalize the grant award process and make sure that grants are awarded based on the merit of the proposal and not the personal relationship between the grant administrator and the inventor.

GOVERNMENT TECHNOLOGIES AVAILABLE

There are databases listing hundreds of technologies that were developed within the various branches of the U.S. government that are now available for the private sector to commercialize. NASA, the U.S. Department of Energy, National Laboratories, all branches of the military, and others offer technologies. Use of many of these technologies is more or less free for the asking. Contact the specific agency that has topics of interest to you, or see the Federal Laboratory Consortium for Technology Transfer at www.federallabs.org

Patent Searching and Patenting

You can perform a patent search at any of the over eighty U.S. Patent Resource Centers referred to previously, or use the following websites for free patent searching.

USPTO

Patent searching online: www.uspto.gov/patents/process/search/index.jsp

Step-by-step tutorial of the entire searching process: www.uspto.gov/video/cbt/ptrcsearching

Printable Patent Searching Guide: www.uspto.gov/products/library/ptdl/services/7_Step_US_Patent_Search_Strategy_Guide_2014.pdf

Process for Obtaining a U.S. Patent: www.uspto.gov/patents/process/index.jsp

GOOGLE PATENTS

www.google.com/patents

FREE PATENTS ONLINE

www.freepatentsonline.com

EUROPEAN PATENTS

www.epo.org

http://ep.espacenet.com

WORLD PATENTS—PATENTSCOPE

www.wipo.int/patentscope/en

For more advanced patent searching and to search in-depth databases, consider the following fee-based services. Also see Research Resources in the next section.

DERWENT WORLD PATENTS INDEX (DWPI)

DWPI is the world's most comprehensive database of enhanced patent documents.

www.thomsonreuters.com/derwent-world-patents-index

PROQUEST DIALOG

ProQuest Dialog is one of the more popular fee-based patent databases. In addition to containing full-text patents dating back to 1971, Dialog offers the ability to do special key-word searches and make cross-references not possible on other databases. It also includes scientific and technical information. Patent law firms often frequent this site.

ProQuest
789 E. Eisenhower Parkway
Ann Arbor, MI 48108

Toll free: 800-521-0600
Phone: 734-761-4700
www.proquest.com

PATENTWIZARD

PatentWizard is a software tool created by patent attorney Michael S. Neustel of Neustel Law Offices, Ltd., to assist inventors in drafting and filing their own "provisional" patent application with the USPTO. It guides you through the steps; you need only fill in the blanks. It's pretty slick.

Neustel Software, Inc.
2534 South University Drive, Suite 2
Fargo, ND 58103

Phone: 888-382-7690
info@neustelsoftware.com
www.patentwizard.com

NOLO PRESS

Offers a variety of do-it-yourself guides for inventors by patent lawyer David Pressman and Richard Stim, Esq., including *Patent It Yourself* (the world's best-selling patenting guide), *The Inventor's Notebook* (documenting your invention), and *Patent Pending in 24 Hours* (filing a provisional patent application. Available at bookstores and direct from the publisher:

Nolo Press
950 Parker Street
Berkeley, CA 94710-2524

Phone: 800-728-3555
Fax: 800-645-0895
customersupport@nolo.com
www.nolo.com

Research Resources

Many of the publications listed here can be found in major libraries, which is the way I prefer to reference some of them, particularly the catalog directories and the Martindale-Hubbell directory. Their Internet sites have searchable databases as well. Some also have searchable CDs available.

U.S. GOVERNMENT PUBLISHING OFFICE

For a wealth of free and inexpensive government publications on just about any subject imaginable, including patents and trademarks.

U.S. Government Publishing Office
732 N. Capitol Street NW
Washington, DC 20401

Toll free: 866-512-1800
contactcenter@gpo.gov
http://bookstore.gpo.gov

SECURITIES AND EXCHANGE COMMISSION (SEC)

The U.S. Securities and Exchange Commission oversees SEC 10-K reports and other information about all publicly held companies. It discloses a company's products, target markets, market position, competitors, sales projections, sales history, complete financial details, and information about lawsuits settled or pending.

SEC Headquarters
100 F Street NE
Washington, DC 20549

Phone: 202-942-8088
help@sec.gov
www.sec.gov

MARTINDALE-HUBBELL LEGAL NETWORK

A database of over one million lawyers and law firms in 160 countries. Use its free Lawyer Locator to find lawyers and law firms by location and/or by practice area.

Martindale-Hubbell
121 Chanlon Road, Suite 110
New Providence, NJ 07974

Toll free: 800-526-4902
Phone: 908-464-6800
www.martindale.com

GOOGLE.COM

Business directories populate the Internet, but when I can't find it anywhere else, I go to Google:

www.google.com

THE DIRECTORY OF MAIL-ORDER CATALOGS

Contains over thirteen thousand mail-order companies selling consumer products throughout the United States. The companies are arranged in over forty chapters by product area. The publisher also offers authoritative reference directories for direct marketing, demographics, business statistics, research, health care, international trade, the food industry, and education.

Grey House Publishing
P.O. Box 56
Amenia, NY 12501

Toll free: 800-562-2139
books@greyhouse.com
www.greyhouse.com

OXBRIDGE COMMUNICATIONS

Publisher of the following four directories, and others pertinent to inventors and marketing:

National Directory of Catalogs: More than twelve thousand North American catalogs including products carried, personnel, list rental data, and production info.
National Directory of Magazines: Over twenty thousand U.S. and Canadian magazines, journals, and tabloids with advertising rates, circulation, and more.
Standard Periodical Directory: Circulation, advertising, and list rental data for more than fifty-nine thousand North American magazines, newsletters, newspapers, and directories.
Oxbridge Directory of Newsletters: Includes over thirteen thousand newsletters, bulletins, and fax letters.

Oxbridge Communications, Inc.
39 W. 29th Street, Suite 301
New York, NY 10001

Toll free: 800-955-0231
Phone: 212-741-0231
info@oxbridge.com
www.oxbridge.com

CHEMICAL ABSTRACTS SERVICE (CAS)

A comprehensive database with over forty million records and referrals to over two hundred other databases in science, technology, business, and patents, administered by their science and technology division.

CAS
2540 Olentangy River Road
Columbus, OH 43202

Toll free: 800-753-4227
Phone: 614-447-3700
help@cas.org
www.cas.org

Small Business Resources

These include both government and private resources geared especially toward helping entrepreneurs and business start-ups. They offer free resources, information, services, and/or assistance.

THE U.S. SMALL BUSINESS ADMINISTRATION (SBA)

Offers free information for budding entrepreneurs. Publications, some of which cost a small fee, may be accessed from the SBA.

SBA
409 3rd Street SW
Washington, DC 20416

Toll free: 800-827-5722
answerdesk@sba.gov
www.sba.gov

SMALL BUSINESS INNOVATION RESEARCH (SBIR) PROGRAM

Administered by the U.S. Small Business Administration, SBIR is a highly competitive program that encourages small businesses to explore their technological potential and provides incentives for them to profit from commercialization. Small businesses must meet certain eligibility criteria to participate in the SBIR program.

- American-owned and independently operated
- For profit
- Principal researcher employed by business
- Company size limited to under five hundred employees

Each year, eleven federal departments and agencies are required by SBIR to reserve a portion of their research and development (R&D) funds for awards to small businesses. The following agencies designate R&D topics and accept proposals: U.S. Departments of Agriculture, Commerce, Defense, Education, Energy, Health and Human Services, Homeland Security, and Transportation; the Environmental Protection Agency; the National Aeronautics and Space Administration; and the National Science Foundation.

Agencies make SBIR awards based on small business qualification, degree of innovation, technical merit, and future market potential. Small businesses that receive awards or grants then begin a three-phase program.

Phase I is the start-up phase with awards of up to $150,000; Phase II awards are up to $1,000,000. Phase III is the period during which Phase II innovation moves from the laboratory into the marketplace. No SBIR funds support this phase. The small business must find funding in the private sector or from other non-SBIR federal agencies.

Small Business Administration
Office of Technology
409 Third Street SW
Washington, DC 20416

Toll free: 800-827-5722
Phone: 202-205-6450
answerdesk@sba.gov
www.sbir.gov

STATE SBIRS AND OTHER STATE PROGRAMS

Some states have their own SBIR programs partially funded on the state level. Contact the website listed above or your state's office of business development for more information about funding on this level.

This may seem enticing; however, remember that SBIR grant solicitations are very specific and the government's requirements and your research will have to match exactly. Some research companies make a business out of acquiring SBIR grants.

A clever inventor with an entrepreneurial bent could approach a business with SBIR experience, propose a joint arrangement, and share in the grant proceeds. Look hard enough and you may find the right match.

SMALL BUSINESS DEVELOPMENT CENTER (SBDC) PROGRAM

The SBA administers the SBDC Program to provide free management assistance to current and prospective small-business owners. SBDC offers one-stop shopping for small businesses by providing a variety of information and guidance in centrally located, easily accessible branch locations.

www.sba.gov/sbdc

SCORE ASSOCIATION (FORMERLY SERVICE CORPS OF RETIRED EXECUTIVES)

The SCORE Association is a national, nonprofit association with more than 13,000 volunteer members and 348 chapters throughout the United States and its territories. SCORE is a resource partner with the U.S. Small Business Administration. Volunteer members provide free individual counseling and business workshops for aspiring entrepreneurs and small business owners.

SCORE volunteers abide by a code of ethics and hold all client discussions as confidential. The volunteer's business specialty and expertise are matched with the needs of the client. Chapter locations can be identified by a city and state search online or by phone.

Of particular value is SCORE's Business Resource Index. This is a comprehensive compendium of good business resources. If you want a one-click business and market research link, try this one. It will directly forward you to many of the business resources referred to in this book.

SCORE Association
1175 Herridon Parkway, Suite 900
Herridon, VA 20170

Toll free: 800-634-0245
Phone: 202-205-6762
help@score.org
www.score.org

YOUR FIRST BUSINESS PLAN

A useful and concise book by Joseph Covello and Brian Hazelgren.

Sourcebooks, Inc.
1935 Brookdale Road, Suite 139
Naperville, IL 60563

Toll free: 800-432-7444
Phone: 630-961-3900
www.sourcebooks.com

Inventor Resources

Some of the best sources of information and support for independent inventors are the local inventors' organizations throughout the United States. Finding them can sometimes be a challenge. Many local inventors' groups do not have a budget for maintaining permanent addresses and phones. As a result, when you look in the phone book or on the Internet under Inventor or Patent Attorney, you may only find patent attorneys and the alleged "rip-off" marketing and submission companies. I trust the following resources, but it's possible for a questionable organization to slip in, so always do your research.

Included in this section is a very diverse mix of nonprofit and for-profit organizations, each with unique resources for inventors.

BY KIDS FOR KIDS (BKFK) 

By Kids For Kids is becoming the world's leading enterprise that connects youth with industry. The company mission is to inspire and motivate youth. BKFK sponsors kids' inventor contests, awards prizes, and offers other opportunities.

By Kids For Kids Co. (BKFK)
1177 High Ridge Road
Stamford, CT 06905

Phone: 203-321-1226
info@bkfk.com
www.bkfk.com

THE CANADIAN INNOVATION CENTRE (CIC)

Canada's leading organization dedicated to assisting inventors and innovative companies.

The Canadian Innovation Centre
Waterloo Research & Technology Park
Accelerator Centre
295 Hagey Boulevard, Suite 15
Waterloo, ON N2L 6R5
Canada

Phone: 519-885-5870
Fax: 519-513-2421
info@innovationcentre.ca
www.innovationcentre.ca

THE CARIK MARKETING GROUP, LLC

Jack Carik and his associates pull from decades of combined experience with selling consumer new products to retail chains, and provide inventors with an objective evaluation of their inventions for a very small fee.

The Carik Marketing Group, LLC
17610 East Kirkwood Drive
Clinton Township, MI 48038

Phone: 248-408-3968
inventorservices@aol.com
www.inventorservices.com

COLLEGIATE INVENTORS COMPETITION 

Operated by Invent Now, Inc., the Collegiate Inventors Competition is a national competition that has awarded cash prizes to the top patentable research since 1990.

Invent Now, Inc.
3701 Highland Park NW
North Canton, OH 44720

Toll Free: 800-968-4332
Phone: 330-762-4463
collegiate@invent.org
www.collegiateinventors.org

DIMWIT'S GUIDE FOR INVENTORS 

A new self-help website for inventors created by Docie Development, LLC. Free information about patenting and commercializing ideas, new products, and technology. My DIMWIT software tool offers a state-of-the-art tutorial to guide inventors and create a presentation of the invention.

www.dimwit.com

DOCIE DEVELOPMENT BLOG ARTICLES 

I have an educational blog about various subjects, including providing services for commission only, getting cash without a patent, maximizing the value of trade shows, submitting ideas confidentially, and Invention Scams Unmasked.

Docie Development, LLC
73 Maplewood Drive
Athens, OH 45701

Phone: 740-594-5200
idea@docie.com
www.docie.com

Invention Marketing Blog: www.inventionmarketing.tumblr.com
Inventor Spot Blog: www.inventorspot.com/writers/ron_docie_sr
How to Work a Trade Show: www.docie.com/faqs/why-should-i-attend-a-trade-show
Invention Scams Unmasked: www.inventionmarketing.tumblr.com/post/1043524359/invention-scams-unmasked

EDISON NATION

Edison Nation is the leading online community and resource dedicated to inventors and idea people. Edison Nation was created by the producers of Everyday Edisons, the invention television series that aired on PBS.

Edison Nation, LLC
520 Elliot Street
Charlotte, NC 28202
Phone: 704-369-7347
questions@edisonnation.com
www.edisonnation.com

THE ENTREPRENEUR NETWORK (TEN)

TEN Online has a very informative website by Ed Zimmer, a self-made entrepreneur and inventor.

The Entrepreneur Network
TEN Online
edzimmer@TENonline.org
www.TENonline.org

INVENTNET

The first Internet-based inventors' organization and the first email-based discussion group. Its goal is to provide independent inventors with information to help them develop and market their inventions. Victor Lavrov is the founder.

InventNET
1519 E. Chapman Avenue, Suite 337
Fullerton, CA 92831
info@inventnet.com
www.inventnet.com

LAMBERT & LAMBERT

Trevor Lambert is one of the more successful invention brokers who will work for commission-only. He also offers an invention evaluation service for a small fee.

Lambert & Lambert, Inc.
11180 Zealand Ave N
Minneapolis, MN 55316-3594
Tel: 651-552-0080
info@lambertinvent.com
www.lambertinvent.com

THE LICENSING EXECUTIVES SOCIETY (LES)

LES, Inc. (United States and Canada) is a professional society comprised of more than four thousand members engaged in the transfer, use, development, management, and marketing of intellectual property. The Licensing Executives Society International has a worldwide membership of more than nine thousand in more than thirty-two national societies, representing over ninety countries.

Licensing Executives Society, Inc.
1120 Route 73, Suite 200
Mount Laurel, NJ 08054
Phone: 856-437-4752
info@les.org
www.lesusacanada.org

INNOCENTIVE

InnoCentive engages in the new and growing field of open innovation. It matches "seekers," companies that are seeking a solution to a problem, with "solvers," any individual or team who may solve the problem. Each problem has a predefined reward. The person who submits the winning solution earns cash awards from $5,000 to $1,000,000.

InnoCentive, Inc.
245 Winter Street, 2nd Floor
Waltham, MA 02451

Toll free: 855-CROWDNOW
Phone: 978-482-3300
www.innocentive.com

THE INNOVATION INSTITUTE

The WIN-I2 Innovation Evaluation Service is an inventor assistance service that uses Preliminary Innovation Evaluation Service (PIES) to provide inventors with an objective third-party analysis of the risks and potential of their inventions.

Innovation Institute
17551 North Old #7
Sturgeon, MO 65284

Phone: 913-522-2674
questions@wini2.com
www.wini2.com

INSPIRE INNOVATION EXPO 

Supported by the Minnesota Inventors Congress. Since 1958, the MIC has annually hosted this world's oldest invention expo in Minnesota. During the event, Inventing Success Workshops help educate inventors about the product development process and how to identify reliable resources. They provide assistance for inventors from any state, throughout the year. For info on the Inspire Expo, go to www.inspireexpo.org.

Inventors Resource Center
P.O. Box 71
Redwood Falls, MN 56283

Toll free: 800-468-3681 (800-INVENT1)
Phone: 507-627-2344
info@inventorsresourcecenter.org
www.inventorsresourcecenter.org

INTELLECTUAL PROPERTY INSURANCE SERVICES CORPORATION (IPISC)

IPISC offers insurance products relating to legal enforcement of intellectual property rights, patent infringement, and the defense of accusations of infringement. Bob Fletcher, Esq., is the president and founder.

Intellectual Property Insurance Services
9720 Bunsen Parkway
Louisville, Kentucky 40299

Toll free: 800-537-7863
Phone: 502-491-1144
info@ patentinsurance.com
www.patentinsurance.com

INVENTION CITY

Mike Marks and Dan Fulford offer their "Brutally Honest Review" of your invention, and the choice few winners will be considered for product development and licensing. Founded in 1998.

Invention City, Inc.
P.O. Box 493
East Orleans, MA 02643

Phone: 320-584-2055
dan@inventioncity.com
www.inventioncity.com

INVENTRIGHT.COM

Stephen Key and Andrew Krauss combine over thirty years experience with marketing and licensing new products to share marketing tips, information, and networking for independent inventors with entrepreneurial tendencies.

InventRight.com

Toll free: 800-701-7993
Phone: 650-793-1477
andrew@inventright.com
www.inventright.com

NATIONAL CONGRESS OF INVENTOR ORGANIZATIONS (NCIO)

NCIO features free education and resources for new inventors including the Inventing 101 online course, special reports, newsletters, and more.

National Congress of Inventor Organizations (NCIO)
8306 Wilshire Boulevard, Suite 391
Beverly Hills, CA 90211

Toll free: 866-466-0253
Phone: 424-278-4928
ncio@inventionconvention.com
www.inventionconvention.com/ncio

NATIONAL INVENTOR FRAUD CENTER, INC.

Owned by Neustel Law Offices, LTD and compiled by patent attorney Michael Neustel, Esq., this is one of the more comprehensive websites for independent inventors.

National Inventor Fraud Center, Inc.
2534 South University Drive, Suite 4
Fargo, ND 58103

Toll free: 800-281-7009
Phone: 701-281-8822
neustel@inventorfraud.com
www.inventorfraud.com/goodguys.htm

NORTHWEST INVENTION CENTER

Features workshops and science demonstrations for kids, hands-on workshops for teachers, and exhibits for museums. Director Ed Sobey, PhD, has written several books on inventing and science for kids, including How to Enter and Win an Invention Contest. A listing of invention contests for all ages is available through the center's website.

Northwest Invention Center
15848 NE 92nd Way
Redmond, WA 98052

Phone: 425-861-3472
ed.sobey@gmail.com
www.invention-center.com

OCEAN TOMO PATENT AUCTION

Ocean Tomo brands itself as the first Intellectual Capital Equity firm. It started the first live, open auction of patents, now held four times per year at various international locations. Technology and high tech are the primary categories of inventions sold. Anyone can submit a patent for consideration. Ocean Tomo also created the nation's first stock market index based on the value of intellectual property. The Ocean Tomo 300 Patent Index launched in 2006 and is priced on the New York Stock Exchange Euronext.

Ocean Tomo, Inc.
200 West Madison, 37th Floor
Chicago, IL 60606

Phone: 312-327-4400
info@oceantomo.com
www.oceantomo.com

QUIRKY

Quirky accepts consumer product ideas from inventors and the public. Winning concepts are funded for product development, and selected finalists are sold in retail outlets. The idea submitters who influence the project share in receiving a percentage of sales revenue.

Quirky, Inc
606 West 28th Street, Floor 7
New York, NY 10001

1-866-5Quirky (1-866-578-4759)
questions@quirky.com
www.quirky.com/shop

TAMARA MONOSOFF

Tamara Monosoff is an inventor, educator, and author of The Mom Inventors Handbook: How to Turn Your Great Idea into the Next Big Thing. Tamara is a public speaker and offers online product launch programs.

Tamara Monosoff

support@tamaramonosoff.com
www.TamaraMonosoff.com

TRADE SHOWS FOR INVENTORS

Periodically there are local and regional inventor conferences and trade shows; however, they have come and gone over the years and few stand out consistently. They tend to be associated with the USPTO Inventors Assistance Center (listed above), which hosts an annual conference for inventors, or by local inventor groups. Refer to the Internet and to organizations listed in this Resources section for current events in your area.

TRIDENT DESIGN, LLC

Chris Hawker and associates are product designers, and they help inventors strategize and create multimedia presentations for potential crowdfunding. They spearheaded the crowdfunding of a cooler prototype that raised over $12 million in fifty-two days.

Trident Design, LLC
939 Burrell Ave
Columbus, OH 43212
614-291-2435
info@trident-design.com
www.trident-design.com

WISCONSIN INNOVATION SERVICE CENTER (WISC)

WISC provides market research and new product assessments for inventors and businesses looking to develop a new product or expand their market share. Operating since 1980, anyone from any state may utilize WISC's services.

Wisconsin Innovation Service Center
University of Wisconsin-Whitewater
1200 Hyland Hall
Whitewater, WI 53190
Phone: 262-472-1365
Fax: 262-472-1600
innovate@uww.edu
www.uww.edu/wisc

Crowdsourcing, Crowdfunding, and Open Innovation

This a continuation of the basic information about crowdfunding and crowdsourcing found in Chapter 3, page 63.

The online encyclopedia Wikipedia, launched in 2001, is a pioneer of what we think of as modern-day crowdsourcing. Now there are many subcategories of crowdsourcing, including crowdfunding, crowdtesting, crowdvoting, crowdcreation, open innovation, eco innovation, and social innovation, to name a few.

Crowdsourcing utilizes the Internet to access online communities and individuals throughout the world. These people are referred to as the "crowd," or more simply, the public. Usually some aspect of crowdsourcing is free, either for the people seeking something from the crowd or for those who provide something to the crowd, or both. Crowdsourcing is the online version of "ask and ye shall receive." What is received spans from money, to knowledge, to opinions, and beyond. The list of uses gets longer every day, with over two thousand crowdsourcing websites worldwide.

CROWDFUNDING

Kickstarter, www.kickstarter.com, figured out that random individuals will donate unthinkable amounts of money to fund the most unique, interesting, and even personal projects. Their success has created a groundswell of what is now called crowdfunding. To understand how these incredible websites work, it's best to simply access some of them and see for yourself. See the list on pages 252–53 for examples.

For inventors, crowdsourcing is a godsend. You may use the crowd to fund your project, establish market interest, test pricing and configuration, you name it. You may invite the crowd to design your brochure, name your product, choose preferred merchandizing schemes, create online communities, and more.

Crowdtesting is an up-and-coming variation for the software field. Over half of all developers now integrate crowdtesting in the process of doing their product development and marketing.

CROWDSOURCING AND OPEN INNOVATION

As an early stage inventor, how might you take advantage of some form of crowdsourcing while keeping your idea secret? Well, you may not keep all aspects of it secret. That's why this is called "open" innovation. But there's good news. Open innovation provides new avenues for inventors to earn money that were never before available! Certainly your patent rights are a traditional way to protect your brilliant ideas, though as we learned earlier in this book, patents have their inherent limitations. They may be ineffective at truly protecting most consumer products as well as many technologies. The trick then is to find a company that will pay you for new ideas, without having to rely on patent rights or the threat of lawsuits. Here is an example of one company that will do just that.

Quirky, Inc., www.quirky.com, founded by Ben Kaufman in 2009, pioneered the funding of inventor's consumer products, and they've gone one step further. They develop, produce and market the new idea at no cost to the inventor. What's the catch? You must have an idea that will sell! Quirky's success stems from its use of the crowd (their community) at nearly every step in the process. They even utilize their community to provide suggestions for product improvement, sales, and other aspects of the process.

When you submit your idea, Quirky asks the crowd to vote for how much they like your idea as well as ways to enhance it. Quirky moves forward on the ideas that are most popular and feasible. Their in-house development team will collaborate to have finalists' products manufactured and offered for sale to retail markets, including Amazon.com, Quirky.com, and stores like Home Depot, Bed Bath & Beyond, Best Buy, and Target. Some stores even have Quirky sections. When the product is sold, Quirky shares a percentage of sales revenue with the idea contributors who had an impact. For some inventors, open innovation is the best show in town.

In yet another open innovation model for paying inventors for their ideas, companies like NineSigma (www.ninesigma.com), IdeaConnection (www.ideaconnection.com), and InnoCentive (www.innocentive.com) act as brokers between organizations that need problems solved (seekers) and the inventors (contributors) that may solve the problem. These brokers are contracted by the seekers to find contributors. The problem may be a very specific scientific challenge, such as a disease-resistant wheat seed, or general ideas about how to improve the seeker's products or markets. In this case, the crowd is you and other contributors. Scientists, engineers, and all types of people from around the world check out the challenges offered on the brokers' websites and choose projects to work on. They typically do so in their spare time from their office or home. If you are one of the contributors whose solutions are accepted by the seeker, then you receive the predetermined reward that was offered by the seeker. It may be $500, $5,000, $50,000 or more. Processes like these are a bit more formal, and all parties sign a contract before work is begun.

Crowdsourcing is exploding into exciting new areas. Here's another example: A snack food company's sales were eroding, so they solicited the crowd with their "Do Us a Flavor" campaign. They received 1.2 million flavor suggestions, and picked six of them. They then asked the crowd to vote on their top flavor pick. After one million votes were received, they paid the winning flavor contributor $75,000 and 1 percent of future sales. Company sales increased 14 percent.

U.S. GOVERNMENT CROWDFUNDING LEGISLATION

The Crowdfund Act was passed by the U.S. Senate on April 5, 2012, as part of the JOBS Act, which allows for the raising of start-up capital on the Internet. An objective of the act is to expand investments from "angel" investors and venture capitalists. Some states have passed their own legislation to do the same. It is thought that these measures, along with approval by the Securities and Exchange Commission (SEC), and support by the president of the United States, will open up the avenue for millions of dollars of new funding for inventors and small businesses.

When you consider all this as a whole, the future couldn't look brighter for independent inventors and entrepreneurs. Here is a small sampling of what is being done with crowdsourcing, crowdfunding and open innovation.

CROWDFUNDING LINKS

The following crowdfunders are just a few examples of the various types of crowdfunding available for inventors, entrepreneurs, and social changers. There are hundreds of other sites to choose from. If you want to launch a fund-raising campaign, you may want to consider enlisting the services of a crowdfunding consultant. The articles and directories listed below will help you find resources. Author's Note: The author assumes no responsibility for the reliability or credibility of any resources referenced herein. Always do your own research to assure the viability of any service provider.

Funding of Ideas, Inventions, and Causes

Kickstarter: the original and more prominent, www.kickstarter.com
Indiegogo: second most prominent and more diverse, www.indiegogo.com
Appbackr: when your project is a new app, www.appbackr.com

Appfunder: when your new app needs development help, www.appsfunder.com
GoFundMe: world's #1 personal fund-raising website, www.gofundme.com
Rocket Hub: funds diverse and personal projects, www.rockethub.com
StartSomeGood: "Dream Big, Raise Funds, Do Good," www.startsomegood.com

Debt and Equity Funding for Projects and Businesses

AngelList: a leading funder of business ventures, www.angel.co
Crowdfunder: a place to invest or raise money, www.crowdfunder.com
Circleup: attract investment for your consumer product, www.circleup.com
Fundable: equity funding of your emerging business, www.fundable.com
Seedinvest: investment for your high-tech start-up, www.seedinvest.com
Microventures: angel and venture capital, www.microventures.com
EquityNet: equity capital from angel investors, www.equitynet.com
Fundingcircle: loans to small businesses, www.fundingcircle.com/us
OnDeck: debt lending for small business, www.ondeck.com

Crowdfunding Services

Trident Design: creates presentations for inventors, www.trident-design.com
Crowdfund Capital Advisors: Sherwood Neiss and Jason Best, www.crowdfundcapitaladvisors.com
Grow VC Group: global funding advisors, http://group.growvc.com/
Peerbackers: crowdfinance advisors, www.peerbackers.com
Donnie Maclurcan: crowdfunding consultant, www.sou.edu/economics/maclurcan.html
Invested In: services for crowdfunders, www.invested.in

Open Innovation for Inventors

Quirky: develops and commercializes products for inventors, www.quirky.com
InnoCentive: post and solve challenges for a prize, www.innocentive.com
Idea Connection: solve problems for monetary prizes, www.ideaconnection.com
NineSigma: open innovation provider, www.ninesigma.com
Innoget: research solutions platform, www.innoget.com
Challenge.Gov: government challenges and crowd sourcing (U.S.), www.challenge.gov
Crowd Spring: crowdsource graphic design and logos, www.crowdspring.com
99Designs: crowdsource graphic design, www.99designs.com
Ideastorm: Dell idea storm process, www.ideastorm.com
Unilever: open innovation for social good, www.unilever.com/innovation/collaborating-with-unilever
GE Ecomagination: submit ideas for environmentally friendly products, www.ge.com/about-us/ecomagination?c=home
Ideaken: collaborative crowdsourcing; seek for solutions and solve problems, www.ideaken.com

Picnic Green Challenge: ideas to save the planet, www.greenchallenge.info
Open Ideo: solve big challenges for social good, www.openideo.com

Crowdsourcing/Open Innovation Articles, Current Topics, and Community

Crowdfunding sites by category: www.ianmack.com/howto-crowdfunding/platforms
Ian MacKenzie's crowdfunding blog and workshops: www.ianmack.com
Chance Barnett's blog, articles, and videos: www.crowdfunder.com/learn
Christian Kreutz's blog: "A space for changemakers," www.wethinq.com
CrowdsUnite: directory of hundreds of sites, www.crowdsunite.com
OnDeck: articles and advice for small business lending, www.ondeck.com/blog
Nesta UK: various resources on succeeding with open innovation, www.nesta.org.uk/develop-your-skills
Crowdfunding Kit: crowdfunding collective, www.crowdfundedkit.com
Massolution: compiles trade reports and statistics, www.crowdsourcing.org
Entrepreneur magazine: crowdfunding articles, www.entrepreneur.com/topic/crowdfunding
Entrepreneur magazine: crowdsourcing articles, www.entrepreneur.com/topic/crowdsourcing
Inc. magazine: crowdfunding articles, www.inc.com/crowdfunding
Inc. magazine: chart, search: "22 Crowdfunding Sites," www.inc.com
Definition of crowdsourcing and helpful links: www.en.wikipedia.org/wiki/crowdsourcing
Building Change Trust: resources for the social innovator, www.buildingchangetrust.org/social-innovation/resources
The Social Innovation Partnership: support for social innovation projects, www.tsip.co.uk
Innovation Community: online innovation community, www.innovation-community.net
Ushahidi: crowdsourcing crisis information, www.ushahidi.com
Root Cause: one-stop shop for social innovation, www.rootcause.org/resources
Humanitarian Innovation Open: innovation challenges to solve humanitarian problems, www.humanitarianinnovation.org
Young Foundation: disruptive social innovation, www.youngfoundation.org/about-us

GLOSSARY

Abstract—A description in a patent that gives a brief (one paragraph) overview of the patent's subject matter. In applications it's located in the back; in issued patents it's on the title page.

Aftermarket—Refers to products and accessories sold for use after the original equipment was sold. (Example: When a battery comes with an automobile, it is part of the original equipment. When you replace a battery, it is referred to as an aftermarket item because you are making a change to the original equipment.) *See also* OEM.

Assignment—Used with intellectual property to denote the sale of an invention. To assign is to sell your patent or other intellectual property rights. You would be the assignor, and the recipient would be the assignee.

Buyer—Refers to that person in a company who is responsible for purchasing goods for that company. (Example: One buyer at Wal-Mart may be responsible for purchasing only automotive accessories, while another buyer is responsible for purchasing batteries and tires.) A buyer for a company is also referred to as a purchasing agent. Not to be confused with consumers who purchase items from stores for personal consumption, called end users. *See also* End user.

Capital—The financial resources and property that a company owns or requires to do business. (Example: The company might need $10 million to purchase the equipment necessary to capitalize a project.)

Cash flow—An accounting term referring to the flow of money both into and out of a given entity. (Example: A positive cash flow means you have more money coming in than going out.)

Chain stores—Basically retail stores with a whole lot of locations, like a mass merchandiser, but they could be a chain of niche market stores. (Example: JoAnn Fabrics has a chain of stores.) *See also* Mass merchandisers.

Champion—A person within a company who likes you and your invention and helps you along with your commercialization endeavors. Also known as an angel.

Claims—That part of a patent that defines your invention. It establishes what it is that your patent gives you the right to exclude others from making, using, selling, offering for sale, or importing. Refer to Appendix A for greater detail on patenting.

Class—Used by the USPTO, it describes any one of over four hundred categories of invention that distinguishes one type of invention from another. (Example: Class 359 is for Optics.)

Co-licensing—When you combine two or more proprietary technologies or patents and package this grouping of intellectual property into one deal or offering. (Example: You and another inventor may join forces to provide an enhanced invention with several patented attributes.)

Commercialize—To transform something into a profitable enterprise. To manufacture items, then distribute and sell them.

Confidential disclosure agreement (CDA)—An agreement between parties whereby at least one of the parties agrees to keep something confidential, such as the invention you intend to disclose to the signing company. CDAs are sometimes referred to as non-disclosure agreements. *See also* Non-disclosure agreement.

Copyright—A right granted by the Registrar of Copyrights in a work of authorship enabling you to keep others from plagiarizing your literary, dramatic, musical, or artistic works. Also referred to as "a fighting chance" in a lawsuit.

Cross license—When Company A wants to use Company B's invention, while Company B wants to simultaneously use Company A's invention. (Example: You may want to incorporate your patented technology into theirs, and receive product back from them that was enhanced with their own patented technology.) This is more commonly done between companies.

Deal—To make a business arrangement, agreement, or enter into a contract. Although it does not refer to the handing out of cards in a poker game, the risk for inventors ever getting a good deal is very similar.

Delegated function—When someone has responsibility for a specific task. (Example: Company N is responsible for manufacturing, Company S is responsible for sales and marketing, and so on.)

Demographics—The measure of a population of people differentiated by their physical characteristics such as age, sex, education, income, and so on. *See also* Psychographics.

Design patent—A patent that is limited to the specific ornamental aspects of a physical object. It doesn't matter how it works, just how it looks. (Example: Tire manufacturers will obtain design patents on a new tread pattern.)

Distribution channel—Refers to the route that a product takes from raw goods or manufacturer to the end user. (Example: After a product is manufactured, it may be sold to a wholesaler, which in turns sells it to a retailer, which in turns sells it to a consumer. This is a three-tier channel of distribution through a wholesaler. Sometimes large retailers will purchase directly from manufacturers and bypass wholesalers; this channel is referred to as manufacturer direct.)

Distribution share—A measure of the scope of a manufacturer's distribution to different wholesalers and/or retailers. Similar to marketshare, but instead of measuring how much the manufacturer sells, the measure is how many places it sells to. (Example: If there are a thousand possible stores that could carry your invention, and Company A and Company B both sell some of their products to all thousand stores, then both companies would have 100 percent of the distribution possible for your product.) *See also* Marketshare.

Distributor—A company whose purpose it is to receive goods from manufacturers and sell them to retailers or others. Manufacturers may also perform the role of distributor especially if they sell directly to the retailer.

Divided territory—When a geographic area is divided so that different companies will cover different regions. (Example: One manufacturer distributes east of the Mississippi, and another to the west.)

Doctrine of Exhaustion of Rights—A law that limits the sources from which you can collect royalties or other fees for your invention. Also applies to collection of infringement damages. Putting it simply, you may generally only collect remuneration from one level of the distribution channel.

End user—The final person(s) who would purchase or use a product or technology. (Example: Customers at a retail store would be the end users if they use rather than resell the products they purchase. The end user of an industrial product may be another manufacturing company.)

Entrepreneur—That person who starts and/or operates a business, usually the founder, president, or owner. Entrepreneurs maintain greater control, ownership, and risk than a hired manager or executive.

Equity—The value or total worth of a business or a property. (Example: If you have a 10 percent interest in a company worth $1 million, your equity interest is $100,000.)

Exclusive license—A license offered by an inventor to only one company to manufacture, use, and/or sell an invention in a given market or territory.

Expediter—A company that distributes a product to a retail store, different from a distributor because an expediter will send a person into the retail store to take inventory and reorder the product. In this respect, expediters are more hands-on and have greater control over the merchandising of the product than distributors.

Field of use—When markets and/or territories are segregated. In a licensing contract, this term can limit the scope of activity. (Example: Company R can sell only to retail outlets and Company O can sell only to organic farmers, and so forth.)

Focus group—A group of people gathered together for the purpose of being interviewed or observed for their opinions or responses about their preferences for a given product. Used in test marketing.

Grace period—A period of time you are allotted to make foreign patent applications and perform certain patent office actions and procedures before you become time barred at the end of such grace period.

Grant—This term is used in two different ways in this book. First, it can mean the award of funds paid to you to develop your invention, such as a federal grant. Grants, unlike loans, usually do not need to be paid back. The second use refers to the transfer of, or right to use, intellectual property. The patent office awards or "grants" you the rights conveyed with a patent. Likewise, when you license your invention to a company, you "grant" it the right to manufacture, sell, and/or use your invention.

Gross profit margin—The difference between your cost of a product or service and the revenue you receive from it. (Example: It costs a manufacturer 50 cents to produce each item, and the company sells it to distributors for 75 cents each. The gross profit margin would be 25 cents, or 33 percent.)

Head fake—To look right and go left, to wear elevator shoes, to say it is "this big" when it is really only "that big." A deception or disappointment.

Independent inventor—An inventor working on their own. Not inventing for an employer or for hire.

Infringe or infringement—To plagiarize or illegally duplicate another's invention. To make, use, sell, or import something that is claimed by another's intellectual property rights.

Initial payment—The first payment you would receive for your invention or other intellectual property. Also referred to as up-front payment, advance, and initial lump sum.

Intellectual property—The legal ownership of patents, trademarks, service marks, copyrights, know-how, trade secrets, and the like.

Interference—When two inventors are claiming the exact same aspect of any given invention. A conflict between two inventors who file the same thing with the USPTO at the same time. (It happens with 1 to 2 percent of all applications.)

Intrapreneurship—A derivative of entrepreneurship where you take the initiative to create and manage a profit center, or separate business, for the company for which you are employed. Instead of forming a company on your own, you form one for your employer.

Jobber—Like a distributor or wholesaler, a company that provides special distribution services.

Job shops—Manufacturing facilities that produce products for others on a subcontract basis. (Example: A manufacturer that temporarily ran out of production space may hire or "job out" another manufacturer to take up the slack.)

Joint venture—Two or more people or entities who form an alliance to reach a common goal. Entities that benefit from a synergistic effort.

Key word—A word you use to perform a search on the Internet or other database that helps you to narrow down the field and find what you are looking for.

Know-how—Knowledge and/or valuable experience that would be of some value to another. Know-how can be part of your intellectual property package; what you know could mean more than who you know, for once.

Least cost manufacturing—When you choose a manufacturer that will produce your product for the "least cost." The combining of orders from several sources in order to get the highest volume discount from one manufacturer.

Letters patent—A way of connoting the transfer of an exclusive right or privilege; another way of saying patent.

Leverage—When you create incentive for another to provide resources and information to you to achieve your goal. To achieve more with less. (Example: An executive provides you with information about an industry for which you have little knowledge. You are leveraging the executive's vast experience with minimal cost to you to obtain such knowledge.)

License—A contractual relationship between two or more parties that conveys rights to intellectual property. (Example: An inventor gives a manufacturer the right to commercialize his/her invention in return for cash, royalties, or other remuneration.)

Licenseability—The degree to which your invention may be a candidate for licensing to another entity.

Licensee—The person or company to whom intellectual property rights have been conveyed. (Example: A manufacturer that acquired the rights to an invention would be the licensee.)

Licensor—The person or company who has conveyed their intellectual property rights to another. (Example: An inventor who licensed an invention to another would be the licensor.)

Litigate—The process of participating in a lawsuit against another.

Mail-order markets—Any arena where you may order a product and have it mailed to you, including ordering from catalogs, websites, direct mail fliers, newspaper advertising, and so on.

Manufacturer direct—*See* Distribution channel.

Manufacturing cost—The amount it costs you to have your product or technology produced by a manufacturer. This term is used on both sides of the manufacturing process. It's the cost the manufacturer pays, and it's also the price you pay a manufacturer for a product. This is an easy term to confuse, depending on where you're coming from.

Market—Also referred to as marketplace. Not the type of market where one may purchase fruits and vegetables; rather those venues where one may sell any given product or service. (Example: Markets where one could market or sell solar collectors would include homeowners, industries, utility companies, and billboard companies.)

Marketability—The degree to which your product or service is perceived to be of value and will sell to the consumers.

Marketing—The process of getting your invention from your hands to a manufacturer, distributor, consumer, etc. (Example: When you are trying to peddle your intellectual property rights to a manufacturer, or selling a finished product to a consumer, the process of distributing and offering your product for sale is the marketing process.)

Marketshare—Refers to that portion of the market that one controls through the sales of a given product or service. (Example: If the total sales for wadgets is $100 million, and ACE Manufacturing's wadget sales total $50 million, then ACE would command a 50 percent share of the market for wadgets.)

Market positioning—The image created in the minds of consumers. Manufacturers develop reputations from their advertising, the types of products and services they offer, and how and where they present them to the public. (Example: think inexpensive chocolate, think Hershey's, any food store, anywhere. Think tape, think 3M, etc.)

Market value—The amount something is worth to those who will eventually purchase it. (Example: The market value for your invention is, in the case of licensing, the amount a manufacturer will pay for your intellectual property rights. The market value for the finished product is the amount the consumers will pay.)

Mass merchandisers—Typically large retail stores that carry common and moderately priced items affordable by the greater mass of the population. (Example: Kmart, Wal-Mart, Target, and other major retail outlets are considered mass merchandisers.)

Merchandising—A function of marketing, merchandising gets into the fine points of how you offer your product for sale, and in what form. How you choose to display your product on a store shelf, and in what combination with other products, the color of the package, and so forth, are functions of merchandising.

Niche market—A specialty market or one with limited scope. Very expensive and unusual items usually have niche markets. (Example: An expensive diamond-studded necklace would not be found at mass merchandisers. You would find such a niche item in an exclusive shop. If your invention is for plumbers only, then plumbing is its market niche.)

Non-disclosure agreement (NDA)—Often used to describe a confidentiality agreement. Instead of saying "confidential," you stipulate that the party will not disclose information. *See also* Confidential disclosure agreement (CDA).

Non-exclusive license—A license offered by an inventor that is not limited to only one company. (Example: An inventor may license an invention to two or three companies to sell in the marketplace, in which case these companies may compete against each other with the same patented products.)

OEM—Refers to original equipment manufacturers. (Example: General Motors is the original manufacturer of equipment such as automobiles and trucks. This is different from companies that produce products for use on motor vehicles after the vehicle leaves the dealership. *See* Aftermarket.)

Online sales—A form of mail-order sales specific to orders placed over the Internet.

Patent—The rights to your invention, granted by the government, to exclude others from making, using, selling, offering for sale, or importing your invention without your permission. Specific patent terms are further described in detail in Chapter 2 and in Appendix A.

Patentability opinion—An opinion rendered by a patent attorney or agent, usually in the form of a written letter, describing your odds of receiving a patent, listing prior art, suggesting the possible scope of your patent, and noting the type of patent you may receive.

Patent agent—A professional registered with the USPTO to file and prosecute patent applications. A patent agent cannot give legal advice or litigate. Patent agents can render patentability opinions.

Patent attorney—A lawyer registered with the USPTO to file and prosecute patent applications. Patent attorneys may offer legal opinions and litigate.

Perceived value—How others perceive your product or invention. What about your invention others like and how much they like it.

Player—A person actively involved in an industry or company, who is especially influential, knowledgeable, and/or keenly interested in championing you.

Positioning—The art of having your company and its products recognized by the marketplace for its unique attributes. (Example: You may choose to distinguish yourself as low price leader, best values, superior service provider, or leader in your area of specialty, such as mufflers (Midas), chocolate (Hershey's), or Internet sales (Amazon).)

Practitioner—Someone registered to practice in a profession. Patent attorneys and agents licensed to practice in the USPTO are patent practitioners.

Prior art—Other inventions or ideas in existence that are similar to, or offer features related to, your invention. (Example: Prior art may include other patents, publications, trade journal articles, and any other form of knowledge either published or otherwise documented.)

Prior art searches—The process of finding prior art in a field; patent searches are included as a part of this. Prior art searches are broader than a patent search only.

Priority date—(This does not mean going to dinner with the FedEx driver.) The formal date on which you make your first patent application. When you first file a patent application, the date of priority is established, which is the beginning of a grace period for filing foreign patent applications. Not to be confused with the conception date, which gives you preference over another inventor in the event of interference.

Product life cycle—The length of time between when a product is introduced to the market and when sales drop off to negligible.

Product line—When several products of a similar nature are sold or offered together. (Example: When an inventor expands to offer several different product variations, rather than one single product.)

Profit center—An area of sales for a product or invention that is particularly profitable. (Example: Instead of dividing the potential sales for your invention by territory, you may look at sales to different markets and identify those that have distinct potential for profitability.)

Proforma—Used in business planning, a projection of anticipated revenues and expenses. May also be a projection of sales and profit centers.

Proprietary property—Intellectual property that is confidential or exclusively owned. (Example: You may claim private ownership to your invention or want your ideas to be confidential information.)

Prototype—A first-of-a-kind working model or likeness of an invention that emulates at least some of what you want it to do.

Provisional patent—A misleading term because there is no such thing as a provisional patent. When you file a provisional application for a patent in the patent office, you are basically reserving the right to, within one year, convert your application into a regular patent application. It buys you some time and allows you to state "patent pending" for a fraction of the cost of a full patent application.

Psychographics—The measure of a population differentiated by attitudes, behaviors, buying patterns, interests, and other nonphysical characteristics. Related to, but different from, demographics. Used to determine the commonality of those who may purchase your product. (Example: You find that people who have strong religious beliefs and prefer watching comedies on television are the perfect target for your product.)

Pull-through sales—The process of creating demand for a product or service. (Example: Rather than simply having a listing in a catalog or having your product sit passively on a store shelf, you can create enthusiasm for your product that causes the consumer to seek it out. When I do book signings, I am pulling through sales for my publisher.)

Purchasing agent—*See* Buyer.

Ramp-up time—The time it takes to get your invention from inception to working model, from working model to finished manufactured product, from product to consumer, or all of the above. Generally refers to the development time necessary prior to commercialization or sale of a product.

Raw material—The materials used to create a product or process. (Example: In an injection molded plastic product, the raw material is the petroleum and chemicals that make up the plastic pellets used in the molding machine.)

Reduced to practice—A term used in the patenting process. Making a working model or filing a patent application are examples of what would constitute "reduced to practice."

Remuneration—Being paid or reimbursed. What you received in the form of payments or other value. Salaries, royalties, consulting fees, product samples, and a warm and cuddly feeling are all forms of remuneration.

Royalty—A payment received based on a commission from sales of your invention or in some other acceptable form of remuneration.

Sales velocity—How quickly and aggressively your product sells. If your product is a great seller compared to other related items, it has good velocity. The opposite of stagnant sales.

Sales volume—A measure of sales, either in terms of the number of units sold, or a dollar amount.

Scope—The scope of the market may be its breadth, the amount of territory, or potential for sales. In patent terms, scope refers to the breadth or reach of the claims in your patent. (Example: A patent with a broad scope may include all light bulbs utilizing electricity as a power source. A patent that limits a light bulb to use only with tungsten filaments would be more limiting in scope.)

SEC 10-K reports—Securities and Exchange Commission reports available for pubic inspection. Publicly held corporations must file information with the SEC on an annual basis describing the nature of their business, risks, and potential for sales. Meant to be used by potential investors. SEC 10-Q's are quarterly reports.

Seed capital—Early stage or initial investment in a new project or company.

Shoe string—A tiny rope inventors use to hold up their confidence, or, what their egos hang on. Also refers to the tiny budget most inventors have to develop their invention.

SIC codes—Standard Industrial Classification, a three- to six-digit code assigned by the U.S. government to all categories (more than one thousand) of products and services. Recently replaced by the North American Industry Classification System (NAICS).

SKU/Stock Keeping Unit—Each different product on a store shelf, or hung on a hook, has an identifying number used for stocking and reordering. (Example: If you can get three colors or sizes of your product placed in a store, you'll have three SKU's there.)

Subclass—The demographic makeup of most inventors, artists, and writers. Just kidding, I think. A classification used by U.S. and foreign patent offices to separate inventions based on their form or function. (Example: A mirror for vehicles may be class 359: optics, and subclass 864: includes adjacent plane and curved mirrors.) There are over 136,000 subclasses in the USPTO.

Subcontract—When the entity that is charged to do work hires someone else to do all or a portion of that work. (Example: You may license a manufacturer to produce your invention, and they may in turn hire another company to build it, or supply parts. They would be "subbing out" the work.)

Sweat equity—A form of capital, namely, your time and effort, used to fund a company or project. (Example: You invest your time in your project without pay. You are basically deferring payment in the hope that the reward will be an invention or company that is worth more as a result of your effort.)

Teachings—Everything your mom didn't teach you is contained in the USPTO, in the form of patents, that is. Patents are meant to teach others how to make or accomplish something. The text and drawings in your patent are the teachings.

Technology transfer—The act of transferring the rights to your invention to another. (Example: This can be achieved by assigning or licensing your rights.)

Test market—The process of determining whether you have a marketable invention, product, or service. The scope of this definition can be very narrow or very broad, from simply asking people what they think about a product to setting up national sales. The scope of a test market is limited only by its budget.

Tiered risk—The levels and amount of risk encountered at the various stages of development; an incremental risk. As development progresses, stakes become higher, investments greater, and risks compounded. Managing by taking the smallest risks first.

Time bar—A time limit or deadline. Inventors are faced with various time limits in which to file their patent applications and other (patent) office actions, or lose their potential for patent rights. Time bars vary by country and by the type of patent for which you apply.

Trademark or service mark—A word, name, symbol, or device that is used to identify goods and services. Its purpose is to indicate their source (manufacturer) and to distinguish them from the goods and services of others.

Trade secrets—Intellectual property of value to someone that is not readily known by others. An elusive critter with its own set of laws governing how you can deal with it.

Trade show—A convention of manufacturers, distributors, buyers, salespeople, press, and other individuals and companies in a specific trade, such as automotive, hardware, housewares, or consumer electronics.

Turn—Not right or left, rather over and over, on a store shelf. A product turn is when a store sells all the products it ordered and orders more; that's one turn.

Turnover rate—(A.K.A., shelf turnover) The frequency at which products are reordered by retailers and distributors. (Example: If a store reorders a dozen of your products five times per year, this rate of turnover is greater than another store that reorders only three times.)

Undivided interest—In the absence of an agreement stating otherwise, when two or more people are listed as inventors on a patent, they have an undivided interest. All parties have equivalent rights; one party can sell all or part of the invention to a third party without the consent or knowledge of the other patent owners.

URL—Technically, a Uniform Resource Locator. In layperson's terms: the address you must type into your computer to access an Internet or website.

USPTO—U.S. Patent and Trademark Office, located in the Washington, D.C. area, specifically Arlington or Crystal City, Virginia. More detailed information about the USPTO may be found in Appendix A.

Utility patent—A type of patent that describes an apparatus and/or method for achieving your desired result. This is the type that one normally thinks of when referring to a patent. Read more about patents in Appendix A. *See also* Design patent.

Validity—A determination as to whether your patent is valid. Patent attorneys offer validity opinions. Ultimately the court makes the final judgment as to whether your patent is valid; the patent office does not. The USPTO issues the patent, but it doesn't guarantee its validity.

Vendor—An entity, such as a manufacturer, that suppliers products or services to another. (Example: If you sell products to a mass merchandiser, you are that merchandiser's vendor.)

Venture capital—Capital invested by an investment company (referred to as venture capitalists or VCs) that specializes in new or higher-risk business ventures. VCs usually prefer to actively participate in a new venture's management and share in its equity.

Wadget—A cross-breed between a gadget and a widget. Originated in Borneo when a cross-licensing deal went south.

Website—A location on the Internet where companies and individuals display information of use to others, such as company profiles, product information, ordering information, databases, and just about everything imaginable, and then some.

Wholesaler—Comparable to distributor by definition. A company that sells or distributes products to retailers or directly to the consumer in high volume or for a price less than what would be found on the retail level.

Working model—(Not Cindy Crawford or Twiggy.) A prototype that actually works, or the next generation of product development after a prototype that works like it's supposed to.

Works-like model—A working model that works like it's supposed to, but doesn't necessarily look like it's supposed to. Opposite of a "looks-like" model, which looks like it should work, but doesn't function properly.

For additional help with terminology, see Analogous Terms on page 5, and U.S. Patent Office online Glossary at www.uspto.gov/main/glossary.

PATENT AND NEW PRODUCT MARKETING WORKBOOK

CONTENTS

INTRODUCTION

THE PURPOSE OF THIS WORKBOOK IS to help you get your invention manufactured and marketed to the greatest number of people, and for you to be rewarded with the fairest profit for your hard work and investment.

This workbook will help you communicate effectively with a variety of industry contacts as you research the market, target potential business partners, and strike a good deal. You do not need to be the expert. The premise here is that you will be leveraging the expertise of people who specialize in commercializing new products and technologies. The process of doing the exercises in this workbook will give you all the information that you need to successfully get your invention to market.

For the sake of example, this workbook will generally follow the consumer product track. However, the market research principles in these exercises apply to most types of products and technology, including scientific, medical, high-tech, industrial, and more.

This workbook takes a step-by-step approach. In Chapter 1, you document what you've done so far and get an overview of where you need to go. Chapters 2, 3, and 4 help you research the market for your invention, identify which manufacturers are best serving that market, and conduct interviews to determine which manufacturers are the best match for your invention. Chapter 5 will guide you through the steps of presenting your invention to companies, and Chapter 6 will help you negotiate the best offer you can get to commercialize your invention.

This workbook can be useful whether you are an inventor in search of royalties or an entrepreneur eager to create the marketing section of a business plan. If the latter is the case, the exercises in Chapters 2, 3, and 4 will help you get answers to important questions about market size, distribution, competition, possible product/technology applications, and other pertinent data.

If you are undecided about whether to license your invention to a company or manufacture it yourself, the process of doing this market research will provide you with the information you need to make the right decision. An added benefit is that the work you perform here will put you on the road to reaching your goal, and you may be at the finish line before you know it.

Who Can Use This Workbook?

I believe that any individual who is optimistic and willing to get back in the saddle after a mortifying rejection, and who has a gift for gab, can perform most of the exercises in this workbook. You don't need to be a salesperson.

Ninety-nine percent of the process of selling your invention is about finding out what you need to know. Bear in mind that as you go through this process, the invention must ultimately sell itself. You can enhance how you present your invention to the industry, but eventually your invention will have to stand on its own (unless you intend to stand in front of every consumer and sell it yourself).

If you can't take constructive criticism—if you are absolutely dead set on what you think your invention is worth, regardless of what the market may tell you—then you will probably have a miserable time with this process and meet with failure. For these exercises to work, you need to be open-minded and optimistic. But please find some comfort in this: I have performed the exercises outlined in this workbook for over a quarter of a century, and *they work.*

With hundreds of inventions and thousands of contacts with companies, I have never had a problem with anyone stealing an idea. I am not guaranteeing that this won't happen to you. However, the more you resist the process described in this workbook, the less likely you will ever get your invention into the hands of the people who can truly help you and would be more than willing to do so.

A Personal Note

As you get further into this workbook, especially when you start contacting key executives in major companies, you may ask yourself, "Why in the world would the vice president of a multinational company bother to give me an interview, much less answer such detailed questions?"

Let me just say this: Most of the communication I do each week is spent cold-calling companies and interviewing top executives. I continue to be amazed by how helpful people are and how surprisingly easy it is to reach top executives in some companies. Many executives are open-minded and willing to listen, especially when they think their company may gain as a result.

It is still my experience that for each ten people I attempt to reach, I will only be able to reach a handful of them on a timely basis, two or three will not want to give me the time of day, and another two or three will end up sharing so much information about their industry that I am trying to figure a way to get off the phone from them! It is altogether possible to make several useless contacts before you get to the final couple of good ones. So, don't be too discouraged in the initial stages.

I cannot overemphasize the value of the information you can obtain from interviewing two or three key executives in your industry. For information you receive absolutely free, others may pay up to $20,000 or more for market research or focus groups, and the information they receive may not even be as effective as the information that you obtain. This is because you will be doing primary research interviewing the people in your industry who make things happen.

I often learn from key executives about barriers in the industry. In a matter of just a few minutes, they may be able to tell me about an obscure government regulation that prohibits certain features on a given invention, or that a similar invention was introduced to the industry before and miserably rejected. I could write a whole book about inventions that didn't make it.

All too often inventors want to think too optimistically about a product—that they know for sure it will not fail, that it will be a hit in the market, and that no one has tried anything like it before. They are unwilling to accept the fact that their dreams may not come true. Since we know that less than 1 percent of new inventions ever make it to the marketplace, you would be wise to uncover any barriers to the possible commercialization of your invention prior to making a substantial investment of time or money.

More positively, this research process may give you a new direction; you may decide you need to research a new or different market, technological design, or other features of your product. Most successful new product introductions have gone through extensive transformations during their development cycle before being introduced to the marketplace.

A Note on Patents

Putting all your eggs into the patent basket prior to learning fully about market barriers and competitive conditions is not necessarily the best plan. Your invention may go through so many changes and improvements that the initial claims of your patent will have little bearing on the final design or technological configuration. This can lead to extra trouble and costs as you file for additional patents.

I must warn you that most patent attorneys will caution you to avoid any type of in-field market research before applying for a patent. The fact is there is risk either way you go. If you are hasty in jumping into the patent process, you may throw a lot of money down the drain. On the other hand, if you don't file soon enough, you may compromise your potential patent position. This is a judgment call that you will need to make. What has largely been missing thus far is a means to assist inventors in obtaining even the most basic market information without revealing their inventions, so they can avoid unnecessary patent costs and heartache. This workbook fills that gap.

This market research and interviewing process is meant to be done quickly and efficiently, and is to be taken very seriously. There is no reason not to finish your entire interviewing process within a one-to-three month period, even if you do it on a very part-time basis, or during lunch breaks.

You can benefit from doing this market research without disclosing your invention, and without signing any disclosure agreements. If you do decide to reveal your ideas, this process will help assure that you trust the right people and reveal what needs to be revealed at the right time.

How to Use This Workbook

The exercises in this workbook are designed to let you fill in the blanks as you get answers to important questions. You will probably want to copy key sections so that you can use them more than once. Expect to make five to seven copies of the market research sections in Chapters 2, 3, and 4. You will also probably need to make two to three copies of Chapters 5 and 6. *Be sure to make a master copy of the workbook before you begin.* Alternatively, you can use a separate notepad or computer to keep track of data; simply use the market research process described in this workbook as a guide.

To obtain computer printable versions of the worksheets in this workbook, go to Ten Speed Press's website at: www.penguinrandomhouse.com and lookup *The Inventor's Bible*. You can download PDFs of all the forms you need.

When performing the market research exercises, you will be collecting the names of many manufacturers and potential licensees. The Distribution Chart, Appendix B, and the Manufacturer/Licensee Selection Worksheet (M/LSW), Appendix C, will prove to be invaluable resources to help you keep track of your prospects. As you collect data from the workbook's exercises, you will transfer some of this data to the Distribution Chart, the M/LSW, or both. These charts will help you summarize your findings, the conclusions of which will then be used in the Review of Final Candidates, Appendix D. Data Transfer points in the exercises will show you when and where to transfer data. Watch for this symbol: ✍.

Pay attention to the sidebars, where I've included some useful information and tips. I highlight points where you may want to take a particular action with this symbol: 🎯. I highlight helpful hints with this symbol: 💡. *The Inventor's Bible* has additional information, some of which is essential reading. If you have not already read *The Inventor's Bible*, I encourage you to do so. At crucial points in the workbook, I will also refer you to pages in the earlier text with this symbol: 📖.

Using Professionals

Although you can perform all the exercises in this book without outside assistance, there will likely be times when consulting a professional would be prudent. Chapter 7 in *The Inventor's Bible* discusses the value of using professionals, which ones to use, and when. You can also refer to the resources section of *The Inventor's Bible* at any time during this process. If you find yourself stuck, or if you just have a question or two, please don't hesitate to take a break and contact one of these professionals so that you may catapult yourself to the next level on this arduous journey of getting your invention to market.

Practice Makes Perfect

It may seem a bit overwhelming when you first review this workbook. How are you going to keep track of over 150 possible questions while interviewing numerous industry experts?

First, you will not ask all the questions of each person with whom you interview. You will carefully select the questions that apply to your situation. Second, planning is crucial. You should be in a continuous process of reviewing what you learned, and then applying that to the preparation for your next step.

If you perform the exercises in this workbook enough times, before you know it the process will flow intuitively. Interviewing people, who you have never met, can actually be fun, and you can make new friends. So, relax and enjoy yourself.

DIMWIT's Guide for Inventors

I developed www.DIMWIT.com, an online program to use hand in hand with this workbook and *The Inventor's Bible*, as a means of helping inventors move forward in the invention process as quickly and painlessly as possible. All you do is answer some simple questions about yourself and your invention (most of which you choose from a drop-down menu), and DIMWIT uses this information to do two things:

- Automatically prepare a complete, professional presentation of your invention, which you can submit to companies or use to communicate with the professionals on your team. You can select whether or not to reveal any confidential information.
- Offer tailored tips and advice based on the status of your project—it's like excerpting relevant sections of *The Inventor's Bible* that are targeted to your specific needs.

Chapter 1

GETTING STARTED

HOW DO YOU BEGIN? First of all, it helps to picture where you're going and the steps that will take you there. The following project checklist will give you an overview of what lies ahead. You can also use it to keep track of your progress. After you've completed all the items on the list, you should know whether you have a commercially viable invention.

The project checklist is followed by a background section that will give you an opportunity to properly document your invention and record information that will be useful for those who may assist you, such as a patent professional, an attorney, or an agent.

Project Checklist

As you do the exercises in this workbook, document your progress by using this checklist. Check off the steps as you complete them.

❑ **DESCRIBE YOUR INVENTION.**
Date completed: ______________________.

❑ **MAKE A ROUGH SKETCH.**
Date completed: ______________________.

❑ **DO A PATENT SEARCH.**
Date completed: ______________________.

❑ **CONSULT WITH A PATENT ATTORNEY.**
Date completed: ______________________.

❑ **DO YOUR IN-FIELD MARKET RESEARCH, FIRST LEVEL.**
Date completed: ______________________.

- ❑ **INTERVIEW DISTRIBUTORS AND SECOND-LEVEL CONTACTS.**
  Date completed: ____________________.

- ❑ **IDENTIFY MANUFACTURERS/POTENTIAL LICENSEES.**
  Date completed: ____________________.

- ❑ **ORDER CATALOGS AND REPORTS FROM MANUFACTURERS.**
  Date completed: ____________________.

- ❑ **INVESTIGATE ALTERNATIVE MARKETS AND APPLICATIONS.**
  Date completed: ____________________.

- ❑ **INTERVIEW MANUFACTURERS, THIRD LEVEL.**
  Date completed: ____________________.

- ❑ **RECHECK ALL RESEARCH INFORMATION.**
  Date completed: ____________________.

- ❑ **FILE A PATENT APPLICATION, IF APPROPRIATE.**
  Date completed: ____________________.

- ❑ **CONTACT MANUFACTURER'S REFERENCES.**
  Date completed: ____________________.

- ❑ **SUBMIT INVENTION TO MANUFACTURER(S).**
  Date completed: ____________________.

- ❑ **FOLLOW UP WITH MANUFACTURER(S) TO WHOM YOU SUBMITTED.**
  Date completed: ____________________.

- ❑ **DOCUMENT MANUFACTURER RESPONSE.**
  Date completed: ____________________.

- ❑ **SEEK PROFESSIONAL ASSISTANCE, IF NEEDED, TO HELP WITH NEGOTIATION.**
  Date completed: ____________________.

- ❑ **CONTACT A CPA (CERTIFIED PUBLIC ACCOUNTANT) FOR TAX CONSIDERATIONS.**
  Date completed: ____________________.

❑ **CONTACT A PATENT ATTORNEY SPECIALIZING IN NEGOTIATING LICENSES.**
Date completed: ____________________.

❑ **DECIDE ON LICENSE/COMMERCIALIZATION/PATENT STRATEGY.**
Date completed: ____________________.

❑ **NEGOTIATE LICENSE OR FOLLOW THE ALTERNATIVE STRATEGY.**
Date completed: ____________________.

❑ **FOLLOW UP ON PROJECT WITH AT LEAST ONE CONTACT PER MONTH.**
Date completed: ____________________.

Documenting Your Invention

Documenting your invention is important to prove date of concept and the ownership of your idea, and to provide information for your patent professional. You can skip this step if you have already appropriately documented your invention. You can move on to patent research or, if you have already filed a patent application, you can move on to Chapter 2. Even if you have a patent, however, this process can be useful for improvements, new designs, and variations. The market research that you perform in this workbook may reveal a need to modify your original designs and concepts. You may need to return to this stage more than once.

BOOKMARK

See pages 32–34 in *The Inventor's Bible.*

ACTION STEP

If you are inventing on an ongoing basis, you may want to purchase a journal for documenting your inventions. You can use the journal to chronicle your progress on different projects and to keep the information that you gather in one convenient place.

INVENTION DESCRIPTION

Describe here all of the features of your invention, and how it works. Describe its benefits and advantages over existing products or technology. Use additional pages as necessary. Have all pages signed, witnessed, and dated. The witness should not be a close relative, but rather a trusted friend, neighbor, or colleague whom you can track down in a few years in the event you need them to serve as a witness at a hearing.

Invention title: ______________________________

Description: ______________________________

____

____

____

____

____

____

____

____

____

Signed by Inventor: ______________________ Date: __________

Witness Signature:

I have read and understand the above. by ______________________

Witness Name: ______________________ Date: __________

Address: ______________________________

Phone: ______________ Email: ______________________

(Optional) Second Witness

I have read and understand the above. by ______________________

Witness Name: ______________________ Date: __________

Address: ______________________________

Phone: ______________ Email: ______________________

ROUGH SKETCH

Draw a rough sketch here, or otherwise attach illustrations, photographs, or other appropriate visual art. This sketch can be very rough because it is only for documenting the conception date or for communicating the essence of your invention to a professional who is working for you on this project. It is important to have this sketch signed and witnessed just as you did in your invention description.

Invention title: ______________________________

Drawing here:

Signed by Inventor: ______________________ Date: __________

Witness Signature:

I have read and understand the above. by ______________________

Witness Name:______________________ Date: __________

Address: ______________________________

Phone: ______________ Email: ______________________

(Optional) Second Witness

I have read and understand the above. by ______________________

Witness Name:______________________ Date: __________

Address: ______________________________

Phone: ______________ Email: ______________________

Patent Research

What is the point of doing patent research? In addition to determining if any other inventions are like yours, looking at patents in your field will reveal which corporations have developed similar technology and give you a lot of information that you can use to your benefit.

HELPFUL HINTS

You may want to do a little initial market research (see Chapter 2) before you do a patent search. Both of these steps are critical, and either one may help you determine in short order that your invention already exists. Luckily, these first two steps are among the easiest and least expensive steps to take.

Even if you already have a patent, similar patents cited by the patent office may hold keys to information that is important to you. You can find "references cited" on the front page of most patents.

Doing It Yourself

Depending on whether you have more time on your hands or more money, you'll either want to perform your own patent search or hire a patent attorney or agent to do a patent search for you.

BOOKMARK

See pages 34–39 of *The Inventor's Bible* for a detailed discussion of how to conduct this stage of research.

Keep on hand copies of patents found in your patent search, as well as a copy of the patentability opinion letter received by your patent attorney or agent. At some point in the future your patent attorney, licensing attorney, licensing agent, or other professional may want to make reference to these search results.

When You Have a Pending or Issued Patent

If your patent is pending, the patent office will send you a list of references cited, which are also referred to as "prior art references." Your attorney or agent should have a copy of this in your file. Always have your own copies of prior art, referenced patents, and patents found in your patent search readily available to review.

Familiarize yourself with all of this material because in all likelihood someone with whom you communicate while doing your market research may already be familiar with some of the inventions that were referenced in the prior art. You need to understand the distinguishing features that set your invention apart.

Again, keep copies of all prior art references cited by the patent office or found elsewhere in your research. You'll need a file space several inches thick to hold this material. Collecting this information from the beginning will save you a lot of grief later when you may need to look up something.

Additional Background Information

At some point you will need to consult with a patent professional about your project. Even if you've already completed the patenting stage and moved on to hiring a licensing attorney or inventor's agent, it will probably be helpful to have some information on hand detailing who's worked on your project and what they have done.

The following worksheets are meant to provide background information for the professionals working for you. Filling out this worksheet is optional; however, it helps you be prepared in the likely event these questions are asked of you in the future by a professional working with your project.

BACKGROUND WORKSHEET

FILL IN THE BLANKS BELOW, AS APPROPRIATE.

I am the sole inventor of this invention: **YES** ___ **NO** ___ If no, answer the following:

I am a co-inventor with the following individuals: (Optional: List their phone or address, and indicate the nature of their involvement.)

____

____

____

Was this invention developed at your place of employment? **YES** ___ **NO** ___

Did you use your employer's resources to develop your invention? **YES** ___ **NO** ___

Do you have an employment contract addressing intellectual property?
YES ___ **NO** ___

If yes to any of these three questions, list employer (optional):

____

Was a patent search performed in the U.S.? **YES** ___ **NO** ___

Internationally? **YES** ___ **NO** ___

Did you receive a patentability opinion Letter? **YES** ___ **NO** ___

Name, phone number, and address of your patent attorney/agent

____

____

____

WHEN YOUR IDEAS ARE UNPATENTED

Has a written description of your invention been witnessed and signed by an unrelated third party? **YES** ___ **NO** ___

Have you filed with the Document Disclosure Program in the U.S. Patent and Trademark Office? **YES** ___ **NO** ___

Date on which the invention was first offered for sale: _______________,
or if not offered, check here: _______________

Date on which you first began using the invention for your own benefit (except for experimental purposes): ____________________, or not used _________

Has your invention been publicly disclosed (in publications, press releases, articles, advertising, and so on)?
YES ___ **NO** ___ If yes, first date _________ How disclosed?

____

Is a patent pending? **YES** ___ **NO** ___ Type (utility, design, provisional):

____

Not patentable _____

WHEN YOU HAVE AN ISSUED PATENT

U.S. Patent No.: ____________________ Date issued: ________________

(Include a copy)

Has a foreign Patent Cooperation Treaty (PCT) application been filed? **YES** __ **NO** __

List foreign patents by country; indicate pending or issued:

____

Have patent rights been assigned to you from someone else? **YES** __ **NO** __

If yes, from whom? ____

Has assignment been recorded in the U.S. Patent and Trademark Office?

YES __ **NO** __ (Include copy of assignment.)

Summarize other patents issued to inventor:

____

____

Was a copyright registered with the Library of Congress? **YES** __ **NO** __

If yes, date issued: ____________________

Trademark filed? **YES** __ **NO** __ Date trademark was issued: ______________

Trademark name:

____

Chapter 2

FIRST LEVEL: MARKET RESEARCH

BOOKMARK

See Chapter 3 in *The Inventor's Bible* for background on how to conduct market and industry research.

A GREAT WAY TO EXAMINE the market for your invention is by following potential distribution channels. In this chapter you will begin the process of backtracking through the channels of distribution, starting with the user end. The following techniques will help you recognize manufacturers active in your product category, learn the distribution channels, and get referrals for your next level of research. You will

- Identify similar, complementary, and/or competitive technologies/products to your invention.
- Identify the current market: size, volume, distribution channels, and alternative methods for reaching the market.
- Identify and get referrals to key individuals—distributors, sales representatives, consultants, and other experts who can advise you about the market for your invention.
- Identify potential manufacturers. Learn about their market position, size, resources, and reputation.

WHY PRODUCT PLACEMENT IS SO IMPORTANT

Generally speaking, from the standpoint of retail store placement, it is hard to get much better than having your product on the shelf of a mass merchandiser such as Wal-Mart. In order for mass merchandisers to maintain high traffic, they offer the best prices, but they also have to offer items that sell at the greatest velocity and provide the best turnover. So by design, only the best-selling products maintain a presence in these stores. In other words, they represent the cream of the crop. It is your job, if your invention deserves to be among them, to prove that it does.

How do you do this? Mass merchandisers try to stick with products that are proven movers. There are two primary ways that these giants do so: First, they accept products with a proven sales performance record. Second, they accept products from vendors with whom they already have a good relationship and who will accept the return of any products that don't sell in the store. By accepting less risk, stores such as Wal-Mart are able to keep margins thinner and stay price competitive. Specialty chains (also called "big boxes"), such as Home Depot, like to have the same great movers as the mass merchandiser—and in fact, with some items they may even have greater sales performance—but overall, they go much deeper into specialty product lines and have less strenuous sales requirements compared to the mass merchandisers.

Your greatest chance of breaking in with your new product into one of these major mass merchandisers or chains is therefore going to be through those companies whose products are currently on the shelves or through a company that is about to open a new account with those stores. First and foremost, you need to check the shelves and determine who currently has market placement.

NOTE: You might be asking how you can tell which companies are about to open new accounts with these stores if all you are doing is looking at existing manufacturers on the shelves. This underscores the importance of delving deeper into the channels of distribution. The information you obtain later in your market research will help to answer this question.

Where to Start

Think about where you would go to find your invention if it were available to the marketplace. If your invention's end user is a consumer, you will need to determine the retail location most appropriate. For a hardware item, you would go to the hardware department of mass merchandisers such as Wal-Mart, Kmart, or Target; to big boxes such as Home Depot or Lowe's; to two or three specialty chains such as ACE and True Value; to one or two regional hardware chain stores; and to independent hardware stores. Once there, you will perform the market research exercises that follow in this workbook.

For inventions with industrial applications, you would go to the factory, warehouse, plant, or other appropriate facility that would be the ultimate user of your invention.

For a scientific invention, you would look again to the end user. The purchaser of an invention with medical applications, for example, may be a physician, hospital purchasing agent, and so forth. In this case, you would most likely start with the people facilitating sales, such as doctors and hospital administrators.

What's on the Shelf

Come up with a list of stores and locations that you think are appropriate. Generally it is not necessary to make appointments ahead of time; however, I try to choose the least busy time at any given location, especially if I want to interview store personnel.

Write down the name of the store that you are visiting, its location, and the date(s) of your visit(s). Answer as many of the following questions as possible, and as appropriate for each location that you visit. For industrial and scientific inventions you will skip some of the questions; however, look for those that apply.

ACTION STEP

You will need a copy of these questions for each location you visit. Make sure you keep a clean master copy. To obtain computer printable PDF versions of this, and other worksheets, go to www.penguinrandomhouse.com and lookup *The Inventor's Bible.*

Name of store/facility: ______________________________

Location: ____________________ **Date of visit:** ____________________

1. **In what department of the store or area of the facility do you find products in your invention category (home decor, seasonal, electronics, and so on)?**

   ____

2. **What is the amount of shelf space allocated to this product category (for example, six feet of shelf space, two square feet, five hooks)?**

   ____

3. **What is the brand representation (amount of shelf space allocated to each manufacturer; (for example, 50 percent Stanley, 30 percent Black and Decker)?**

   ____

4. **What are the types of materials or manufacturing used by the manufacturers? (Stanley products are all molded plastic, Black and Decker uses primarily metal, and so on.)**

   ____

5. **Price range by manufacturer/brand (Stanley $2 to 5, Black and Decker $4 to 10, and so on):**

   ____

6. **Brand image (Stanley, ease of use and low price; Black and Decker, quality and unique features, and so on):**

____

____

7. **Which manufacturers/brands most closely match the market position suitable for your invention? List your top choices:**

DATA TRANSFER

Also record the answers to this question on the Distribution Chart in Appendix B.

_____________________ _____________________

_____________________ _____________________

_____________________ _____________________

_____________________ _____________________

8. **List the contact information for all manufacturers within this product category. This information is generally on the packaging or box, or the store manager may provide it. (Use additional paper as necessary to document several brands.)**

Brand name: _____________________

Company name: _____________________

Parent company name: _____________________

Address: _____________________

Toll-free number: _____________________

Other phone number: _____________________

Email address: _____________________

Website: _____________________

Patent numbers: _____________________

Country where manufactured: _____________________

NOTE: The amount of market presence, or market share, established by the manufacturers at the locations you are visiting may ultimately be the single most important factor in dictating your commercialization strategy. After going through this process, you may find that certain manufacturers have a strong hold on your product category and control the majority of the distribution to all the sales outlets. In such cases, you may find the choice facing you is to either license your invention to one of these marketing giants or miss out on the majority of the potential market. The good news is that with one manufacturer, you could command distribution to thousands of locations. The downside is that in the event your invention is not accepted by the main manufacturer, you're out of luck at thousands of locations. Watch out for this scenario and keep it in perspective as you go through the remaining market research process.

ACTION STEP

Pick up or ask for the manufacturers' catalogs, brochures, price sheets, and any other available descriptive material.

On-Location Interviewing

Once you have developed a feel for the products in your category at each location that you visit, it is time to start the interviewing process with a knowledgeable and experienced store clerk, manager, purchasing agent, plant manager, or other person best able to give you information about potential customers for your invention. Interview individuals using the questions that follow. Write down their contact info.

ACTION STEP

Review all the questions in this section of the workbook before interviewing and put a star by those that are of particular importance to you. Leave a blank when questions are not appropriate or when the interviewee is not responsive. As you ask questions, try to capture the essence of, or the most pertinent points, in the responses that you receive.

Name of interviewee: ______________________ Date: ____________

9. **What kind of people buy products in this category? (This may vary from location to location. Look for general categories, such as mostly female or male users, people with discretionary income, do-it-yourselfers, tradespeople, teenagers, and so on.)**

____

____

10. What appears to drive sales? (Examples may include price, time-saving or other special features, color, and so on.)

Primary feature: ______________________ Secondary feature: ______________________

Other: ______________________ Other: ______________________

11. What price ranges and features are most popular?

Primary: ______________________ Secondary: ______________________

Other: ______________________ Other: ______________________

12. What is the breakdown of market share by sales volume? (Stanley outsells Black and Decker by three to one, or Stanley 50 percent, Black and Decker 20 percent, and so forth.)

____

13. Market changes: Are sales seasonal, dependent on economic climate, affected by recent changes in ownership of either the present company or the manufacturer, or anything else?

____

14. What are other important factors? List any side comments relating to what may affect sales:

______________________ ______________________

______________________ ______________________

______________________ ______________________

______________________ ______________________

15. What are the names and phone numbers of manufacturer representatives or salespeople who call on this store?

______________________ ______________________

______________________ ______________________

16. Which manufacturers offer better quality, choices, service after the sale, and so on?

DATA TRANSFER

Also record the answer(s) to this question on the Distribution Chart in Appendix B.

Manufacturer: ______________________ Strength: ______________________

Manufacturer: ______________________ Strength: ______________________

17. Which manufacturers have had problems with filling orders, poor product quality, returns, and so on?

Manufacturer: ______________________ Problem: ______________________

Manufacturer: ______________________ Problem: ______________________

18. How many stores or locations are there, and where are they (in the U.S., in the Midwest, and so on)?

DATA TRANSFER

Also record the answer(s) to this question on the Distribution Chart in Appendix B.

No. of locations: _______ Territory served: ______________________________

19. Who supplies the store (its own warehouse, an independent distributor, an importer)? Include names of suppliers and contact information.

DATA TRANSFER

Also record the answer(s) to this question on the Distribution Chart in Appendix B.

____

____

20. Has this person ever seen or heard of a product or technology with features that perform a similar function or achieve the same results as your invention, but in a different way? (Talk about generalities without revealing the invention unless it is your decision to do so.)

____

HELPFUL HINT

What to reveal—Choosing how much to reveal about your invention can be a complex decision. You must factor in the advice of your patent attorney, the results of your patent research, whether you intend to file for patents internationally, the historical practices toward independent inventors by the manufacturers in your industry, and so on. When in doubt, simply describe how your invention may benefit the consumer without revealing how it does it. This way the person you are interviewing can better help you, and yet you are not revealing the secrets of your invention. The response of a store manager or sales clerk after seeing your invention can often prove to be invaluable. This person may have many years of experience in your product category. Again, you must decide whether the risk of revealing your invention outweighs the potential benefit of the information that you may receive from the person you are interviewing. With the following questions you do not need to reveal your invention. For more information about getting specific feedback on your invention, see Chapter 3.

21. If a manufacturer were to introduce a new item along the lines of your invention, which manufacturer might that be? (This is possibly the most important question of all.)

DATA TRANSFER

Also record the answer(s) to this question in column B on the M/LSW in Appendix C.

1. ______________________________
2. ______________________________
3. ______________________________
4. ______________________________

22. What is the standard price markup for this category (50 percent markup, keystone [two times], and so on)?

____

23. What is the turnover, or reorder rate, in your product category, both highest and lowest acceptable range? (Does the store reorder a dozen every quarter? Every other month?)

High: ______________________________

Lowest acceptable: ______________________________

24. Identify collateral sales in this category. (These may be either a company catalog or website of the company you are visiting.)

____

____

25. What are other locations that have similar products/technologies in this category?

Name: ________________ Location: ________________

Name: ________________ Location: ________________

Name: ________________ Location: ________________

ACTION STEP

The major trade shows and conferences in your industry can be excellent places to conduct research. Now is a good time to ask questions about these events. For a guide to this process, refer to Trade Show Information in Appendix A of this workbook.

26. Can you be referred to another employee or person who is particularly knowledgeable about this product category?

Name: ________________ Name: ________________

Company: ________________ Company: ________________

Expertise: ________________ Expertise: ________________

Phone: ________________ Phone: ________________

Other contact: ________________ Other contact: ________________

27. How might you recontact the interviewee?

Name: ________________

Title: ________________

Phone: ________________

Email: ________________

Address: ________________

____

____

ACTION STEP

Now that you have collected some basic information from your market research in the field, you can start to build the Distribution Chart located in Appendix B. In this way you can create an easy and quick visual reference showing which manufacturers serve which distributors and retail outlets. As you continue in your market research, you may learn more details that will help you to complete this distribution chart.

Chapter 3

SECOND LEVEL: INTERVIEWING EXPERTS

THE NEXT STEP IN YOUR RESEARCH is to interview wholesalers, distributors, manufacturer representatives, salespeople, and other industry experts.

Your objectives will be to

- Discover industry trends.
- Explore alternative applications for your invention.
- Define the features of your invention most preferred by the industry.
- Uncover barriers to introducing products/technology in your category.
- Learn more about the manufacturers and their market positions.
- Learn about competition.

Your initial market research should have provided you with the names and locations of manufacturers and with leads to industry experts such as salespeople, wholesalers, and trade associations. Now is the time to contact them. If you did not get specific contact information during your initial interviewing, there are other ways to locate these experts. The Internet, of course, is an invaluable tool. Simply go to a search engine such as Google and plug in key words to find websites for companies and trade associations. If you have the name of a sales representative who works independently, you might find it easier to recontact the person you first interviewed and ask for more specific contact information.

What are you looking for?

It is not enough to just find a company willing to manufacture your invention. Lots of failing and incompetent companies hastily look for new products for their salvation. The key is to identify those companies that will do the most effective job of introducing your invention to the marketplace. Then you must determine which of these companies will work with outside inventors like you.

As you delve deeper into the distribution channel for your product, the people you interview will know more about the nitty-gritty details of your industry, as well as the people and companies that influence the industry. If your initial interviews provided you with a long list of potential manufacturers for your invention, this next series of interviews will help narrow your search. You will be seeking to obtain more details about the market positions of different companies, company strengths and weaknesses, and other important information. All of this will influence your decision about which manufacturers to approach first with your invention, and in what manner.

Another objective of this second stage of interviewing is to explore other possible applications and uses for your invention that are yet undiscovered. Here, you can begin to discover which manufacturers are more appropriate for different configurations of your invention (if you have more than one).

ACTION STEP

Review the answers to your first-level questions in Chapter 2. If important questions were not answered, add these to the following list of second-level questions.

Interviewing Techniques

Interviews are usually done over the phone at this stage, although if you are in a large metropolitan area, you may find it easy to go to a distributorship or visit a salesperson in your area.

Start by explaining that you were referred to them by "Mr./Ms. _____" who suggested that you contact them because they are very knowledgeable about your subject area. (It never hurts to prop an ego along the way.) If this is a cold call, explain that you are an inventor who is attempting to find the right manufacturer for your invention and thought that their wealth of knowledge in this industry might be of benefit. Explain that all you want is to get a referral to whatever company they think you should be dealing with. It is important to make it clear up front that your primary mission is simply to receive a referral. You want to give the impression that this phone call may not take much time. It is also important for them to understand that this is not a sales call.

Once you receive a positive response, move directly into your questioning. Avoid saying that you are doing some market research and would like to take a few moments to ask them some questions. This probably will not bring the desired response. Instead, go directly to your questions.

Your line of questioning will consist of two waves. At first, you are simply trying to get a referral to the right manufacturer. This is especially critical in the event that your interviewee has only a minute to spend with you. In this case, explain that you have an invention relating to such-and-such and that you are trying to find a manufacturer who may be interested in licensing or acquiring the rights to your invention. Document this response in the appropriate questions below.

If the person you are talking with has time to talk further, he or she might ask for more detailed information about your invention. This is a very good sign and allows you to initiate the second set of questions. Either start with the questions that immediately follow, or if you choose to reveal your invention, start with the questions in Getting Specific Feedback about Your Invention, later in this chapter.

Second Level Interviewing

Prioritize your questions before these interviews and stick with those that would be most appropriate to your particular situation. Some of these questions may be particularly important to help you fill in the blanks on your distribution chart.

Name of interviewee: ______________________________

Title: ______________________________

Location of interview: ______________________________

Date: ____________________

28. **If this were his or her invention, which manufacturers would he or she prefer to work with and why?**

DATA TRANSFER

Also record answers to this question in column B on the M/LSW in appendix C.

____

____

____

29. **What are the pros and cons of dealing with those companies?**

____

____

30. **How does each manufacturer position itself in the market? (Which is a low-price leader, high-quality manufacturer, and so forth?)**

____

____

31. What types of accounts or markets are strongest for these manufacturers? (Which is stronger with small retailers, mass merchandisers, or industrial accounts?)

32. Where do these manufacturers have the strongest market presence, by territory, and in which sales outlets?

DATA TRANSFER

Also record answers to this question on the Distribution Chart in Appendix B.

33. Are there rivalries between these manufacturers? Which ones?

34. Which manufacturers have a reputation for unscrupulous practices, either from working with inventors or from the way that they deal with others in their industry?

35. Which manufacturers have a better presence in foreign markets? Which countries?

36. Which foreign manufacturers are making an aggressive impact in the United States? At which sales outlets?

37. What trends tend to affect this industry?

____

____

38. What trends are apparent for products in this category?

____

____

39. What barriers might there be in marketing a product in this category?

____

____

40. What does this person think about the features and benefits of your invention? Are these features and benefits desired by the industry? (Again, you can reveal as much about your invention as appropriate, without necessarily revealing your invention.)

____

____

41. What features and benefits are driving sales in this product category?

____

____

42. How does the price range affect sales, and what are realistic price ranges for a product that offers benefits and features similar to those of your invention?

____

____

43. How important and effective is pull-through marketing for products in this category, and what techniques are used?

____

____

44. Are there other experts in this field who would be good for you to contact? (These may include authors, engineers, professors, consultants, government agencies, or other salespeople.)

____

____

45. Can this person recommend someone at a particular manufacturer for you to contact directly? (Be sure to get the name, title, and contact information for this person.)

____

____

46. What trade publications and directories will best inform you about this industry? Does he or she have a back issue they can share with you?

____

____

____

47. What are the trade associations that serve this industry? Include contact information.

____

____

____

48. What trade shows serve this industry? Include contact information.

DATA TRANSFER

Record answer(s) here and/or on the Trade Show Listing Chart in Appendix A.

____

____

____

49. What are the prominent mail-order catalogs for this product category?

50. What are the URLs of the prominent websites offering online sales in this product category?

51. Is there anything else you should know?

Interviewing for Industrial/Scientific Inventions

Here are additional questions for industrial/scientific inventions and technology. These questions are intended to be asked of plant managers, purchasing agents, administrators, and so on.

52. How will your invention be accepted by engineers, plant managers, physicians, and so on? What are its strong points?

53. Who makes buying recommendations? What's this person's title? To whom does he or she report?

54. Who makes the buying decision? What's his or her title? To whom does he or she report? Which department is it?

55. Which standards or testing are necessary to meet approval? (Example: OSHA, EPA, ANSI, FDA.)

56. How would your invention affect the company's operations? (Example: Would it necessitate a change in procedures, purchase of additional equipment, or layoffs?)

____

____

57. How would your invention affect the company's profit? Would it add value?

____

____

Getting Specific Feedback about Your Invention

If at any point during this interview, you find that your interviewee is particularly knowledgeable about the subject of your invention, it may be beneficial to share more information. If you feel comfortable about revealing the invention itself at any point during this process, here are some considerations:

First, there is generally nothing as important as the "*wow*" factor, as in, "wow, why didn't I think of that?" Once the interviewee understands what your invention is or what it does, you should stop, look, and listen. Do you notice excitement? Are there more questions or confusion?

The degree to which you score high on the wow scale will dictate the ease with which you may flow through this process. The bigger the wow, the more willing your interviewee will be to refer you up to the next level, to another industry expert or a powerful manufacturing contact, and so on.

What to show? If you are conducting the interview in person, a working model or a miniature model may be ideal. In other cases, a photo, illustration, or even a brief verbal description is all that is needed. If appropriate, you can get into great detail about your invention, such as preferences in color, size, configuration, variations for other applications, product extensions, improvements, and pricing. If your interviewee is at some distance, you may send any of the above materials, and/or a video, CD, or Internet presentation.

Particularly for unpatented inventions, it certainly doesn't hurt to have your interviewee sign a confidentiality agreement before disclosing your invention to them if they're willing, or at least ask them to keep your information in confidence.

Here are some questions you can ask:

58. **Has this person ever seen anything like your invention? What, where, and when?**

59. **If so, who manufacturers it?**

60. **What are the price and prominent features of similar products/technologies?**

61. **What does he or she like most about your invention?**

62. **Which of your invention's features and benefits will be most desired by consumers?**

Be sure to satisfy yourself that the people you are interviewing really understand your invention. If you are not sure, offer additional explanation.

After the sales representative or expert you are interviewing knows more specifically what your invention is, and/or does, you should ask the following more specific questions. (Don't worry about repeating yourself.)

63. **Which manufacturers would be most likely to want this product in their product line?**

DATA TRANSFER

Also include these answers in column B on the M/LSW in Appendix C.

64. What price range makes sense for this product?

____

65. Where should it go in the store?

____

66. Any there any drawbacks, areas of concern, or known obstacles?

____

____

67. How many of these products would he or she initially buy (or envision selling)? In what configuration (5-watt model, 12-ounce size, and so on)?

Configuration 1st preference: ____________ Initial order: ____________

Configuration 2nd preference: ____________ Initial order: ____________

Configuration 3rd preference: ____________ Initial order: ____________

68. At what cost would he or she be willing to purchase it from a manufacturer?

____

Completing the Interview

Whenever an interview ends, you should thank the person immensely and send a follow-up note of thank-you by email or mail. Be sure you get the correct name and address of the person with whom you've been speaking.

Name: __________________________

Title: __________________________

Address: __________________________

____

Phone: __________________________

Email: __________________________

Date of interview: ____________________

Questions for Trade Associations

Trade associations can be an excellent source of information. Contact the prominent trade associations referred to by your interviewees. Interviewing an association executive is very similar to any second-level interview, except that your questions should be limited to information specific to manufacturer choices and market trends.

Trade associations are one of the best sources for obtaining information about industry trade shows since these trade associations either sponsor trade shows or are in some way closely associated with them. Find out what trade shows are upcoming.

BOOKMARK

See pages 66–74 in *The Inventor's Bible* for further ideas about how to obtain information through trade associations and trade shows.

In the event that you are lucky enough to have an opportunity to attend one of your industry's regional, and preferably national, trade shows, most of your second and third levels of market research can be done all at one time at one of these shows. It is certainly appropriate at any time during your project to attend a trade show. See Appendix A on Trade Show Information.

DATA TRANSFER

Record information about trade shows on the Trade Show Listing Chart in Appendix A.

Chapter 4

THIRD LEVEL: QUALIFYING MANUFACTURERS

FROM THE MARKET RESEARCH you already performed, you should have a laundry list of potential manufacturers from which to choose. In this third level of market research, you will obtain more-detailed background information about the potential manufacturers/licensees and further judge which of these may be the best candidate to manufacture/license your invention.

Even if you aren't sure whether you want to license your invention or start your own company, what you find out at this stage should help you in making a decision. In the event that you ultimately choose to start your own entrepreneurial effort, the information gathered here will be invaluable for the marketing section of your business plan.

BOOKMARK

See Chapter 4 in *The Inventor's Bible* for background on qualifying manufacturer and licensee candidates.

Bear in mind that throughout this process, you need not necessarily reveal the trade secrets about your invention in order to gain valuable market information. Although this level of market research involves contacting manufacturers directly for the first time, you are not necessarily offering your invention for sale to them. You are still doing research and performing these next exercises to determine to whom you may ultimately want to offer your invention for sale, and how you should go about doing it.

Your objectives at this stage are to

- Obtain catalogues, annual reports, SEC 10-Ks, and other information from prospective manufacturers/licensees.
- Identify manufacturers whose methods of producing products best complement the requirements of your invention.
- Identify companies that have a market position that best aligns with your needs.
- Identify companies whose product line will be best complemented by your product/technology.
- Determine the procedure for the submission of outside ideas, the criteria for invention submission, disclosure agreements, and the manufacturer's review process.

UNDERSTANDING THE MANUFACTURER'S PERSPECTIVE

In recent years, product design teams have taken on much of the business of product development for major manufacturers. These design teams primarily rely on consumer market research to determine consumer needs and develop ideas to solve those needs.

The design team will generate many new ideas and concepts and hone in on those that tend to have the strongest perceived need by the consumer. From this, a manufacturer develops product features which it designs into its product line and into its advertising.

An example of how this works is the Arctica refrigerator line by General Electric, which helped propel GE's market share. Many of the refrigerator features were developed in focus groups, where there were pages of fresh new ideas for innovative products that would keep the company in new product development for the next few years. Basically, through such focus groups, consumers provide feedback and become the innovators.

Since many companies utilize these sophisticated techniques to obtain market research feedback in the process of developing their products, it is understandable why many companies are often lukewarm about inventions presented to them from outside inventors. They know that in most cases the outside inventor does not have the resources to do the kind of market research described above. If they have a choice of developing ten different salable products generated from within the company, versus working with ten outside inventors, none of whom have done a comparable level of market research, you can understand their tendency to want to stick with those concepts developed internally. Not to mention that the ten outside invention projects all require negotiation of some sort of license just to acquire the invention and start the project. Plus, there is the associated cost of attorney fees and royalties, so again, you can imagine how a product manager would prefer to work with ten ideas generated from within the company. The product introduction is simpler, faster, and more secure, and it carries less risk than those ideas coming from the outside.

The bottom line is that for your invention to be accepted, it must be exceptionally revolutionary and already recognized as filling needs that a company has identified from customer feedback. In addition, people in the company, including engineers, product developers, salespeople, and even executives, must embrace these new ideas. All of these people will have their own favorite ideas, thus presenting a possible quagmire.

Luckily for the independent inventor, most small- to medium-size companies cannot afford sophisticated market research like that described above. Therefore, if you end up dealing with a smaller company, the information that you gathered from your market research may actually be more detailed than what your prospective manufacturer/licensee would ever gain independently. As you consider which questions are best to ask of your prospective manufacturing candidate, be particularly mindful of the company's resources and determine how your contributions will fill in the gaps.

Preparing for the Interview

Contact the corporate headquarters of each potential manufacturer/licensee and, when they are a publicly held company, request copies of the most recent Security and Exchange Commission (SEC) 10-K (annual) report and all subsequent 10-Q (quarterly) reports. Then contact the marketing department and order brochures and catalogs representing the company's entire product line. In some cases companies only supply catalogs to their vendors, and in this case you can ask for product flyers or separate catalog sheets that represent that portion of the product line that fits your invention category. The company's website may be very informative as well, but it may not represent the total scope of the company's activities.

BOOKMARK

Before you contact a key decision maker, thoroughly review the discussion on pages 86–88 of *The Inventor's Bible,* under "Picking the Lineup" and "Tricks of the Trade."

Take special note of the company's statement of its overall philosophy, marketing philosophy, the marketing objectives in the SEC reports, and the overall product offering to determine how well your product fits within the company's product line. This is also a good way to look at the features of existing products, many of which you may have not seen thus far in your research. Determine how the features of your invention will enhance what the company has to offer.

After you have done this with each manufacturer that you are considering, you may want to refine your list of candidates, or the order in which you choose to contact companies.

What to Disclose and When

The process of disclosing your invention to someone during your first two levels of market research does not pose the same risks as doing so in level three. Prematurely disclosing your invention to the manufacturers themselves has its own significant set of risks. Even though you may have shown your invention to a distributor or retailer for the purpose of getting feedback, now is the time to be particularly cautious about who you reveal your invention to, and how much you disclose.

Here are some general guidelines about what you should disclose about your invention, when, and to whom. This decision is more of an art than it is a science. In many respects, I believe that how a person feels intuitively can be as important as any other element in making these decisions. Don't be afraid to follow your instincts during this process.

WHEN YOU HAVE AN IDEA ONLY, OR A PATENT PENDING

Except for the communication you have with your patent attorney, agent, or broker, you should not yet disclose any exact working knowledge or trade secrets about your invention. Instead, speak about its features and benefits. The exception to this is if you decide you want a specific opinion about your invention from the person you are interviewing, and you are willing to take the associated calculated risk.

HELPFUL HINTS

You need to find out whether the companies you are approaching will communicate with you without any disclosure agreement in place. If they want a disclosure agreement, will they sign yours, or must you sign theirs? My first preference is for a company to sign my disclosure agreement. My next preference is to sign nothing at all or to just send a cover letter requesting confidentiality. If a company is willing to look at the invention, and if it meets my other qualifications, it is my experience that submitting the invention is better than not sending it at all. For further information on disclosure agreements, see the section at the end of this chapter.

WHEN YOU HAVE A PATENT

I like to disclose information on a need-to-know basis, and therefore I don't like to even disclose the patent number or patent to anyone except those whose market and product feedback may be particularly valuable, and when revealing this information will not compromise a potential business dealing with another manufacturer/licensee.

NOTE: Until you are much further along, you will likely be best served to not formally offer your invention for sale. Consider stating on all correspondence, and in all communication, that "this is not an offer to sell this technology/product/invention."

Questions for the Interview

Some of the questions below may seem redundant. It does not hurt to ask a similar question in a different way. Plus, you may be asking one group of questions at one point in time, and another group of questions at another point in time. Therefore, these questions may overlap, but they are no less important.

Even though your apparent primary purpose is to gather market research information, your ultimate purpose may be to get the company that you are contacting so interested in your invention that it makes you a

BOOKMARK

Familiarize yourself with the discussion in *The Inventor's Bible* on "Making Contact" and "What Decision Makers Can Tell You" on pages 90–97; pay special attention to pages 91–94, "The First Phone Call," which gives tips on how to get the conversation started.

wonderful offer. In essence, you may have dual purposes and yet still not know at this stage which companies you ultimately want to pitch your invention to. Therefore, it doesn't hurt to begin to establish some perceived value for your invention by the person with whom you are dealing as you gather market information.

In most instances you can accomplish this when you talk about the general features or benefits of your invention during your initial interview. If you say nothing more than, "my invention will cost 30 percent less than competitive products," such a simple statement can start to establish value.

The objective is to have this key decision maker start to visualize how your invention can benefit the company, and how it can fit into the company's existing product line. Depending on the amount of time this person has, you may only be able to ask the first group of questions in the initial conversation, saving the second and third group of questions for follow-up calls.

ACTION STEP

Remember, you will need a copy of these questions for each person you interview. Keep a master copy.

The Interview

Use the information gathered thus far to complete the following contact information for your records.

Company name: ________________________________

Parent company: ________________________________

Brand names: ________________________________

Phone numbers: ________________________________

____

Address: ________________________________

____

____

Contact person and position: ________________________________

____

____

Email: ________________________________

Website: ________________________________

Start with these questions. Highlight the three or four of these questions that are most important to you, in case your interview time becomes shortened.

69. What does the person you're interviewing think about the invention concept? (Ask this question to make sure your contacts understand your concept, and what stands out about it. Hint: I can talk about an improved blindspot mirror for vehicles with a 50 percent greater view, without disclosing how my invention works.)

____

____

____

70. Would an invention like this support the company's market position?

____

____

71. Where would an invention like this fit into its product line? (In which product category, with which other items, in which division of the company, and so on?)

____

____

72. What drives sales (price, features, packaging, and so on) in this product category?

____

____

73. What are the ideal sales outlets (mass merchandisers, specialty stores, mail order, and so on)?

____

____

74. How strong is the company in these sales outlets, and in which ones?

____

____

75. At which trade shows does the company exhibit?

____

____

____

76. What is the general retail price range for a product like this? (Get a ballpark figure: $5 to $10, under $1, $100 to $150.)

____

____

77. What is the potential sales volume for a product like this? (Get a broad ballpark figure here—5,000 to 10,000 units annually, 10 to 50 million annually, so at least you know how many zeros are in the person's mind.)

____

____

78. Where do they see the sales? (Nationally, regionally, internationally? Which countries?)

____

____

79. What might be the product life cycle (three years max in its current form, over ten years, indefinite)?

____

____

80. What would be the time frame for introducing your invention? How long would it take to get it to the broader market and how long would it take to ramp up sales (for example, 5,000 the first year, 20,000 the second year, 100,000 the third year, and so on)?

____

____

Decision Point

At this point in the interview, you should have a firm idea as to whether the person you are interviewing sees a great potential value for a new product in this category or not. If he or she seems really sold on the idea of your invention, and especially if time is limited, this is a good point to ask the person if he or she would like to review your invention in greater detail. You can go straight to asking questions about the best way to deliver additional information about your invention.

81. **Is there interest in learning more about the invention, and if so, what do they want to see: a photo, a copy of your patent, a JPEG, an email file, or a working model?**

 ____

 ____

82. **Would they like you to present the invention in person? When, and will they pay for travel expenses?**

 ____

83. **Which address should this be sent to and to whose attention?**

 ____

 ____

84. **How much time is needed to review it and make an initial determination?**

 ____

85. **Who will be reviewing the invention; who would negotiate the final arrangements?**

 ____

86. **If the company needs to return your material, is there a deadline, or a preferred shipping method?**

 ____

At this point, it may not hurt to pull back and, if you don't already know, ask for further information about the company.

87. What experience does the company have working with outside inventors? (For example, how many outside inventors have they worked with in the past, and when?)

DATA TRANSFER

Also list companies that give positive responses to the above question in column C on the M/LSW in Appendix C.

88. Has the company ever paid inventors for unpatented inventions? (The answer to this is particularly important for learning how much emphasis you should be putting on your patent rights.)

89. What royalty rates are typical in this industry and/or for this company (for example, 5 percent of net sales; 2 percent to 3 percent, paid by number of units only)?

90. Does the company typically prefer to purchase patents outright, and does it ever make advance payments?

Important Notes

You will probably find that the person is either happily answering your questions or starting to become a bit annoyed or short with you as you ask questions about the company's practices. You have probably gone far enough to determine whether or not this company may be a good fit. This would be a good time to conclude the conversation by talking about disclosure agreements if you haven't already done so and if you need to. (See the section on disclosure agreements at the end of this chapter.)

It may be in your best interest to avoid making any commitments at this point. Avoid locking yourself into action. If you interview three or four companies and find that one in particular is extremely qualified, extremely interested, has the perfect market position, and would be interested in an exclusive arrangement only, it may hurt your position to have disclosed your invention to other companies. Therefore, you don't want to make a strong commitment to submit your invention to any single company until after you've finished this entire interviewing process, or are extremely satisfied and confident that making a firm commitment is the right thing to do.

In the event that your interviewee is highly receptive and has the time to answer a few more questions, ask the next set of questions, or wait until your next telephone conversation to ask them. Even if you decide this company is not for you, getting answers to these questions will be helpful. If you end up working with a different company, you will have that much more information about that company's competition; certainly, this information may be extremely valuable to the company with which you end up doing business.

Here are some follow-up questions. Again highlight and use those below that are most important to you.

HELPFUL HINTS

If you told an executive that you may send information about your invention and later decide not to do so, it is acceptable and certainly courteous to notify this person that you have changed your mind and for strategic reasons will not be submitting your invention after all. Thank this person for his or her time and consideration.

91. Does this company ever pay consultant fees to inventors?

____

92. Has the company maintained exclusive licenses, and what levels of minimum performance are anticipated in this category?

____

93. Has the company ever had to relinquish a license because it did not meet minimum performance, and why?

____

94. Has the company ever paid to patent inventions for inventors or paid patent maintenance fees?

____

95. Is the company willing to offer the names and phone numbers of outside inventors who have done business with it in the past? List them.

____

96. What experience does this company have with manufacturing products similar to yours? Where are such products manufactured?

____

97. **What experience and resources does the company have for engineering, quality control, and service after the sale? What does the company perceive are its strengths?**

____

98. **How would the company go about determining the manufacturing cost? (Will it need a prototype sent to the manufacturing facility? Are engineering drawings necessary? Is a file sent via the Internet sufficient? What file type?)**

____

99. **Who would be responsible for supplying material to help determine the manufacturing cost? Is the inventor expected to provide engineering drawings or a sample unit? Will the company provide these things?**

____

100. **How finished do drawings or working models need to be? (Is the company looking for a "looks like" model, or a "works like" model, or both?)**

____

101. **Does the company generally do design improvements and refinement in-house or expect this solely from the inventor, or does it prefer to do a collaboration between the two?**

____

102. **Is the company willing to obtain a quote for the manufacturing cost? At what unit volume, and from which source(s)?**

____

Filling in the Blanks

You will undoubtedly not be able to get all your questions answered in your initial interview. In fact, you may only have one shot at a good conversation with a key decision maker. However, there will be other opportunities to get your questions answered. For example, you may start out by dealing with a vice president; he or she will make a strategic decision to accept your invention for further consideration and then suggest that you speak with a product development manager, engineer, or another person in the company. By interviewing different people, you can find out more. Therefore, it does not hurt to repeat some of the previous strategic questions, especially if you have any questions about the initial answers you received.

Bail-out Questions

The protocol for bailing out of an interview is quite simple. At any point during your interview if you find that there is a lack of interest on the part of the interviewee, if you decide that you don't want to do business with this company, or if the interviewee wants to discontinue the conversation, you should always thank the person very kindly for his or her time. Then before either one of you has a chance to take a breath, ask at least one of the following questions. The questions are listed below in order of importance.

103. What other companies would be in a better position to help you, or in a better market position?

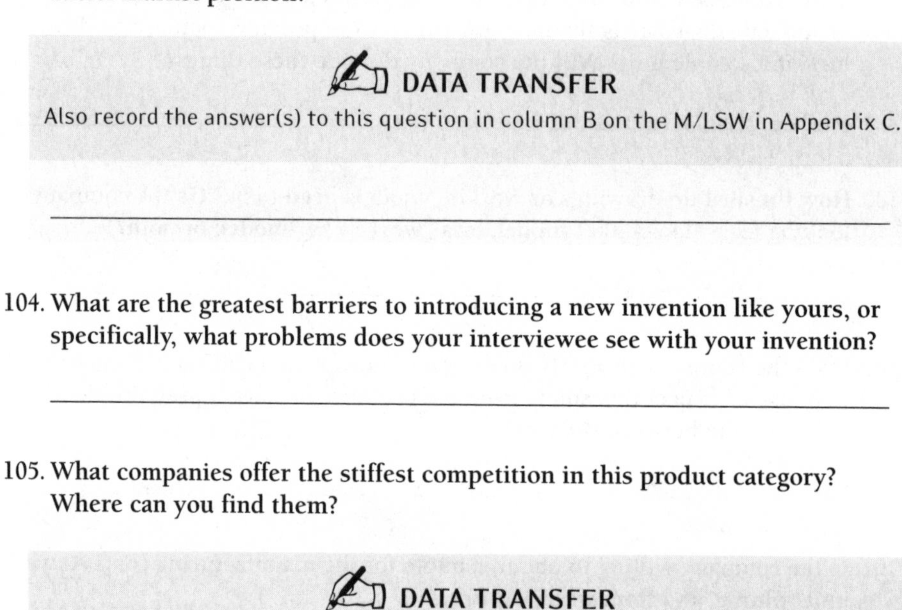

DATA TRANSFER

Also record the answer(s) to this question in column B on the M/LSW in Appendix C.

____

104. What are the greatest barriers to introducing a new invention like yours, or specifically, what problems does your interviewee see with your invention?

____

105. What companies offer the stiffest competition in this product category? Where can you find them?

DATA TRANSFER

Also record the answer(s) to this question in column B on the M/LSW in Appendix C.

____

Disclosure Agreements (What to Submit and When)

As you gather information about manufacturers, you will come across their policies regarding disclosure. I usually advise my clients to sign a company's disclosure agreement if they are unyielding and will not offer any other choices. In my experience, with hundreds of invention submissions to corporations, I have never experienced any company taking advantage of this or publicly disclosing my client's invention.

It causes me little concern. My greater concern is to get the company to review the invention initially. The best way to avoid obstacles is to follow the company's standard operating procedure.

BOOKMARK

Pages 97–104 in *The Inventor's Bible* provide particularly valuable information about what information to initially submit, and whether you should do so under a disclosure agreement or not.

MANUFACTURER TRENDS: A HISTORICAL PERSPECTIVE

During the industrial revolution, as large corporations became multinational, they became more protective about accepting outside inventions and the NIH (not invented here) syndrome was prevalent. If the invention was "not invented here," the company was not interested in it. However, after the recessions of the 1980s, globalization, and foreign manufacturing, research and development departments rapidly shrank in many U.S. manufacturers.

The 1990s began to see greater receptivity toward contributions by outside inventors, even by major, multinational companies. Although many companies still prefer to rely on new product ideas coming from their sales representatives, distributors, and customers, many have become receptive to receiving outside ideas from independent inventors. Companies that once boasted the NIH syndrome now have special departments and even websites dedicated to assisting with the submission of outside ideas. This is the good news.

The bad news is that the concept of confidentiality is almost a thing of the past. In other words, hundreds of companies are willing to consider outside ideas, and virtually none are willing to agree to keep those ideas confidential. It's not that these companies are interested in blabbing to the world information about inventions submitted to them. Even if a company dislikes an idea, why would it want to give any ideas to the competition?

The simple fact is that some companies have had trouble with inventors claiming that the company stole an invention when indeed they did not. It is not uncommon for companies to be developing an invention, or have it submitted to them by another individual, prior to or around the same time that you are submitting your invention to them. No one wants to believe that someone else in the world thought up the same invention; the knee-jerk reaction is to sue the company because surely this company "stole" the idea. Even if the company is not guilty, it may be in the company's best interest to pay damages, and some companies in fact have, rather than go through a lawsuit or damage their reputations. Obviously, companies want to avoid this scenario if they can, and their attorneys have figured a pretty good way to do this: to simply refuse to agree to confidentiality from the start. End of story, end of conflict.

Further, I do not see this as a highly detrimental issue for inventors. In my own disclosure agreement, I do not emphasize confidentiality. Instead, I emphasize that the invention remains the property of the inventor. If a company signs my agreement, this is what it is agreeing to. I see this as much more beneficial than obtaining confidentiality.

If confidentiality is a particular concern to you, then I suggest that you establish at least a patent-pending status prior to the submission of any information to a company. Further, you may choose to submit only brief information about the attributes of your invention initially, and if the company shows keen interest, it then may be willing to negotiate a special confidentiality agreement with you. **NOTE:** The latter may also entail your agreeing to keep the company's communication with you confidential; the company may have more to lose from the information you gain, than vice versa. The issue of confidentiality is one point in the process where your invention project may become stalled, and it is crucial to move beyond this stage.

Remember, you do not want to disclose your invention to any more companies than is necessary. Let's say that you have narrowed down your list to five companies to which you might submit your invention. You may find that three of these companies have worked with outside inventors, whereas two of them, in their fifty-year histories, have never paid royalties to an outside inventor. What would you do? I would immediately rule out the latter two, and only consider the three companies that have honored the contributions of outside inventors. I see this as much more significant criteria than their specific legal stances on confidentiality. Again, this emphasizes the importance of the prequalification procedure performed in these exercises.

ACTION STEP

After you have interviewed all the appropriate manufacturers, refer back to the M/LSW in Appendix C. In light of what you have now learned, cross out those companies who you believe are no longer appropriate candidates and add to the list of other companies you would like to contact. List your semifinalists in column D of the worksheet.

Warning Regarding Disclosure

If you disclose or submit your idea or non-patented invention to a company without having an agreement of confidentiality in place, then such a submission may be considered a public disclosure—even if you did so privately and exclusively. This may be particularly so with large and/or publicly held companies. Such a disclosure may compromise your ability to ever receive foreign patent rights and could prematurely start your one-year grace period to file a patent application. Refer to the Public Disclosure of Information section on pages 33 to 34 for more detailed information. As always, it is best to consult a patent attorney about this issue.

Chapter 5

PRESENTING YOUR CONCEPT TO COMPANIES

BOOKMARK

See Chapter 5 in *The Inventor's Bible* for background on how to formulate a submission strategy.

BY NOW, YOUR INITIAL market research should be complete, and you should be satisfied that you're on the right track. The in-depth market research that you've done will provide you the material you need to make your next set of decisions. At this stage, you'll actually present your invention, or some abbreviated description of it, to prospective manufacturers/licensees. One of the primary objectives even at this submission stage is to receive feedback about your invention. You are not yet offering it for sale. This will come later.

In this chapter you will

- Decide to whom you will submit your invention, with what information, and in what order.
- Learn more about possible barriers and risk factors adversely affecting your invention.
- Learn more about the potential markets and applications for your invention.
- Discover which companies are most interested in your invention and why.
- Develop a strategy for proceeding toward possible negotiation, or use the acquired information for the marketing section of your business plan.

First, take a look at the information you have at this point. From the knowledge you gained during your initial market research and in your interviews with key executives in the industry, you should have a good list of prospective manufacturers/licensees. If you have done your research thoroughly, your list should be narrowed down to one to five companies or so. Rarely are there more than half a dozen companies in any given product or technology category that assume a given market position. Even if your list is longer than this, the following process will help to further narrow down the field.

Reviewing Manufacturer Candidates

Use the following checklist to further evaluate each of your manufacturing/licensing candidates.

Company Name: ______________________________

A. Does the company's market position closely fit your invention? (Examples: A high-quality $50 item would not fit well with a company whose entire product line consists of products that sell for less than $10. A U.S. manufacturer that would be hard-pressed to manufacture your invention for less than $10 would be a weak fit if foreign manufacturers can produce it for $2.)

❑ **YES** ❑ **NO**

B. Do the company's strengths complement your needs? (A company that has few resources for product development or engineering will not be a good fit if your project requires more extensive engineering and product development. On the other hand, if the company can assure you of a good minimum sales performance, you may be justified in hiring engineering support to compensate for the weakness in this area.)

❑ **YES** ❑ **NO**

C. Does the company seem keenly interested in your invention and has it expressed a willingness to extend the resources necessary to take the invention to the next stage in a timely manner?

❑ **YES** ❑ **NO**

D. Is the company willing to consider product extensions, to produce variations for multiple applications, and to meet the total market potential for your invention?

❑ **YES** ❑ **NO**

E. Does the company have a positive track record working with outside inventors and reimbursing them fairly?

❑ **YES** ❑ **NO**

F. Have you satisfied yourself that you are submitting your invention to the appropriate person in the company, or the appropriate division?

❑ **YES** ❑ **NO**

G. Have you determined exactly what information the company is looking for, and are you prepared to supply this?

❑ **YES** ❑ **NO**

H. Are you satisfied that you are in an appropriate position with regard to your patent strategy to reveal your invention at this time?

❑ YES ❑ NO

I. Do you have a good feeling about working with this company?

❑ YES ❑ NO

For answers marked no, decide whether such a deficiency is crucially adverse to your project. If it is critical, ascertain if it is practical to overcome the deficiency, and what steps you should take.

ACTION STEP

After concluding this review process, rank your final candidates in your order of preference and list them in column E of the M/LSW in Appendix C.

If you are satisfied with the answers to these questions, along with any others that may be pertinent to your specific project, then your next step is to start the presentation process.

HELPFUL HINT

Meeting in person—In the event you present your invention in person to a company, I advise that you be accompanied by a professional negotiator, business agent, or attorney. If nothing else, they can serve as witness to discussions and be sure to take detailed notes.

Submissions in Writing

My standard operating procedure is to start with only my number one manufacturer/licensee candidate, whether or not the invention is patented. If the response is negative, then I proceed to the next best target, and so on. If time is of the essence, after submitting your invention to your number one or two picks, you may want to approach two or more companies at a time. Remember, regardless of whether it is positive or negative, the response from any company may help you to enhance your presentation to the following company. The main thing is to get an initial response within one to two months, and the sooner the better.

BOOKMARK

Review pages 113–132 in *The Inventor's Bible* for more information about submitting your invention, and specifically to pages 115–117 for "What to Submit" and a sample submission letter.

PRESENTATION WORKSHEET

It helps to keep a record of your submissions. Use the following worksheet to keep track. (Make a copy for each company to which you present your invention. Make sure also to keep copies or photos of the material you submit.)

Company name: ______________________________

Contact name: ______________ Phone number: ______________

Email: ______________ Fax: ______________

Shipped to address: ______________________________

____

____

Submission date: ______________ Shipping method: ______________

Other specific instructions: ______________________________

____

Detailed list of materials sent: ______________________________

Date of follow-up call (allow five to seven days): ______________

Date review to be completed: ______________

Names of others involved in review process:______________________________

____

Date of response (if rejected): ______________ Date material returned: ______________

Questions for Follow-up Calls

After having submitted your material to a company, the next step is to learn what the company thinks about your invention and whether it is interested in pursuing the invention further.

The best, and possibly only, way to get a quick response is for you to recontact the person in the company with whom you are dealing and receive a verbal response. It is not uncommon for a person to initially seem particularly eager to see your invention and then sit on it for two or three months after having received the information from you. This is because new product development from an outside source is often last on a company's list of priorities. Marketing their existing products, introducing new products already waiting in the wings, and handling administrative concerns nearly always take priority over new ideas coming from outside.

As a result, you need to make the extra effort to contact people and keep trying until you get them. This will make the difference in whether this stage of the process is three months long or nine months long. In many of my successful licensing deals, I have had to provide the impetus to maintain a necessary level of communication to get the deal done.

Before making this call, review the list of questions you used in your initial interview with this company. Make a note of questions that were not answered or were not answered completely. You can use this second interview to start to fill in the blanks and learn more about the company's strengths, the importance it places on patents, its experience with new product introduction, and so forth.

HELPFUL HINT

Even if you get a rejection letter, you should still go through the following process of making a follow-up call. Often an initial rejection changes to acceptance and an eventual deal.

If the company is truly interested in learning more about or obtaining the rights to your invention, you should notice a willingness on their part to pitch the attributes of the company to you. If you are dealing with someone high in the company who does not have the time to get into these types of details, ask if he or she will recommend someone else who can provide some details about the company so that you may learn more about them and in the course of doing so, enhance a possible working relationship.

Follow-up Interview

Within seven to fourteen days after the company receives the material from you, start by initiating the following follow-up interview, even if the company review process is not finished or has not yet started. It is important to make sure that the appropriate person received the information. You should make sure that your invention is clear to this person and that he or she has all the necessary information. You can use this opportunity to ascertain how soon the company will be finished with the review so that you will know when to call again. Make sure that you diligently make the second call around the time that the review is to be finished.

Date of follow-up interview: ____________________ **Name of person you recontacted:** ____________________

106. What is the overall impression of the invention?

____

____

107. Was the information you sent clear, and did it thoroughly explain the advantages and benefits of your invention?

---

108. What additional information may the company need to better understand the invention?

---

109. Is the company interested in considering your invention further?

IF THE RESPONSE IS NEGATIVE

If the response you get is negative, you can still use this call to get information that may help you when you submit your invention elsewhere.

Basically you want to know why it was rejected, and this isn't necessarily a bad thing to ask outright. However, in order to get effective answers, you will likely need to ask more specific questions, such as the following:

110. Has the company seen anything like this invention? What is the closest thing to it?

---

111. Does the invention fit the company's market or support its market position?

---

112. Are there problems with a product like this? What are they?

---

113. Are there known barriers in the market for introducing this invention?

---

114. Are there possible manufacturing problems or other production complications?

____

____

115. Has the company ever attempted to market a similar product, or have other companies attempted a similar concept?

____

____

116. Is the projected price an obstacle? If so, what price range would make the invention acceptable for reconsideration?

____

117. Did the company obtain a manufacturing cost quote? What was the cost and from where was it obtained?

____

118. Is it simply an issue of bad timing, and is there another time in the near future when would be better to resubmit the invention? To whom should you submit it?

____

119. Has there ever been a need expressed in the marketplace for an invention like this?

____

____

120. Are there possible modifications (such as different size or other features) that would make the invention more acceptable?

____

____

121. Are there applications for your invention in other markets? Which markets?

____

____

122. What other companies may be in a better position to commercialize the invention? (Get city and state if possible.)

DATA TRANSFER

Also record the answers(s) to this question in either column B or F on the M/LSW in Appendix C.

___

___

123. Are there industry experts who may provide additional beneficial information? Where to find them?

___

___

WHEN ALL RESPONSES ARE NEGATIVE

When the responses you receive from each and every company are all negative, you need to determine why. Are there insurmountable barriers in the industry? Do you need to redesign your invention to get the cost down or make it less complicated? Is it a good invention for smaller niche markets but it lacks enough sales potential for major distribution? Are you satisfied that the companies to which you presented the invention represent all the possible product applications?

If you believe the barriers to be insurmountable, at least you know that you tried.

The experience of rejection can be quite grueling. For example, I have presented an invention in one market, only to find that there was better market potential in a market that I had not yet identified. Then I would go through the entire process in a second market only to find out that there was a third potential market application and an entirely different set of companies to contact yet again. Each time I had to take the same steps.

Once you have done this two or three times, only to receive negative responses, you may have to give up hope. Then again, sometimes it's a matter of timing. Maybe you'll be able to try again in the future and be surprised with a positive response.

WHEN YOU RECEIVE A POSITIVE RESPONSE

Anything that is not an outright rejection of your invention you should take as a positive response. Pay attention to specifics. For example, does the company want to do more investigation, does it want to know on what terms you are making the invention available, and so on?

Be ready for your contact to ask, "Are you working with any other manufacturer?" or, "Is there anyone else looking at this invention?" At this point, considering neither of you has made a commitment, you are under no obligation to discuss this subject. It would be a true statement to simply say, "We have not offered the invention for sale to any other company at this time."

The fact is, even if you have presented your invention to a second company, you are still in the process of prequalifying companies. You haven't offered it for sale. At this point, you're still involved in market research. This underscores the importance of stating in all submissions, "this is not an offer for sale."

Your objectives at this stage are to determine: Is the company interested? How much is it interested? What resources and commitment might it commit to your project, and on what terms? This is all important information to know for when you start your negotiations.

Respondent's name and title: ______________________________

Date of response: ______________ **Their location:** ______________

124. What was the response? (Document it in detail. Quotes are helpful.)

____

____

____

125. What is it about the features and benefits that is most appealing? (Again, get into specifics here.)

____

____

____

126. Does the company need additional information to enhance further investigation, and what does it need?

____

____

127. What is the internal procedure for further review?

____

128. Which departments will handle the review, and who specifically will be doing the review?

____

129. Who reports to whom? Who makes the final decision?

____

130. When will the review be complete?

____

131. Is it okay to contact the people doing the review to provide further clarification?

____

Defining the Terms

It is not uncommon for a company to want to know what you want for your invention, your expectations, and your level of reasonableness. Naming a specific price is probably not a good plan at this point. You want to appear reasonable and flexible. Here is what you might say:

"I am really not sure what my invention is worth. I've researched this market and determined that your company is a/the top competitor in this market. You probably will have the best idea about the potential market for this invention, and I trust you as the experts in this field. I was hoping to hear what you think you can do with this invention over time. I am more than willing to be reasonable about whatever terms you think would be fair for an invention of this nature. I want terms that you are comfortable with."

As you can see from the above, you have completely defused any notion that you are an unreasonable inventor or have unreasonable expectations. Even if you want a million dollars for your invention, the point is to get the company to commit to what they think they can do. If company A says it thinks that this is a $10 million market, company B thinks it is a $5 million market, and company C thinks it is a $30 million market, then this is revealing information about which company is going to be a better candidate. These projections will be largely contingent on the breadth of distribution the company commands; a company with excellent distribution potential is in a better position to provide you with a higher figure.

Ask the following questions:

132. What kind of sales volume do you envision for this product over time and at what sales price?

____

133. Has the company estimated a manufacturing cost, and from what source(s)?

____

____

134. Is the company more interested in an outright purchase or in paying ongoing royalties?

____

135. Is the company interested in an exclusive arrangement? In which markets? For which applications?

____

136. For how long is the company interested in an exclusive (for example, five years, ten years, or for the life of patent)?

____

137. What foreign markets does this company see for your invention and in which of these markets does the company have a stronghold or good distribution?

____

____

Finally, if it is your intention to attempt to conclude a licensing deal with this company, ask these next critical questions:

138. What will be the deciding factor(s) that will determine whether you want to proceed with this project? (For example, if the sales department embraces it, the company will go for it. Engineering must approve it, and then it goes straight to the owner of the company for final consideration.)

____

____

139. **What is the timeline for the final determination?**

____

140. **Is there anything that you, the inventor, can do in the meantime to support or positively influence the final decision?**

____

____

141. **What is your contact's gut feeling about the chances of this project proceeding and why?**

____

____

Post-Presentation

Beyond this point, I recommend prayer. You need a champion within the company, or you will have little chance of succeeding. You now must rely on your champion to ride herd on the company. At some point the invention has to stand on its own, and you are almost at that point. It is like sending a child out into the world after graduation. He or she has to make it on his or her own.

This is where the "wow" factor comes in. If your invention can stand on its own and impress people, and you have a passionate champion within the company, you will probably enjoy pretty smooth sailing. However, you still have work to do. You still have the next stage, negotiation, with its own associated risks and factors that can prove fatal to your project, even when your experience has been all positive up through this stage. It is my experience that when deals have gone smoothly up to this stage, they still have only a 50 to 75 percent chance of final completion. This is not the time to let your guard down.

If you receive an extremely positive response from the company and they want to proceed further, the next chapter about negotiation will help you along.

Last Chance to Bail

Short of actually negotiating with a company, this may be your last real chance to obtain good detailed information about the company's stance in the market, its strengths and weaknesses. In the event that you eventually choose to start an entrepreneurial effort and manufacture and market the invention on your own, getting to this stage has given you tremendous information about the market and your eventual competitors. You may realize that you will have no significant competitors because you've just interviewed the majority of your potential competitors and now know that they have no interest in proceeding further along this product line. You may have an excellent product and technology and none of the companies serving that field appreciate its value.

At minimum you have rich information to include in the marketing section of your business plan and will be able to share some confidence with your investors that there is little likelihood of any strong competition from major players in the field—at least initially. This can be an excellent selling point when you pitch your business plan to venture capitalists or others. It is very seldom that start-ups obtain this kind of behind-the-scenes knowledge of their competitors, and investors should be impressed by this.

Public Disclosure

It is important to note that in many instances you can get all the way through this step without revealing your trade secrets, and certainly without making a public disclosure of your invention. If so, you may keep alive the opportunity to file both U.S. and foreign patents. It can be a distinct advantage to use the tactic of delaying the filing of a patent application since companies, with whom you do business, will sometimes take over the patenting process and pay for it too.

Always consult with your patent professional when forming such strategies because there are risks inherent with delaying the filing of your patent application.

Chapter 6

NEGOTIATING THE BEST DEAL

BOOKMARK

Read Chapter 6 of *The Inventor's Bible*. Be sure to familiarize yourself with all the different terms that typically go into licensing contracts. In order to improve your chances for a successful negotiation, you will need to understand the terms that are most important to the company.

CONGRATULATIONS! You are finally at the negotiation stage. This is when you can really appreciate all your hard work, when you'll be able to use all the research that you've gathered.

There are many details to consider at this point, and this chapter will help make them less overwhelming. It will help you

- Evaluate the final licensing candidates.
- Determine whether you should hire a professional negotiator.
- Develop a checklist of important factors and considerations to review before negotiation.
- Review the key terms to be negotiated, weigh positives and negatives, and make your final decisions.

In order to help you determine an appropriate negotiation strategy, start by reviewing all the information you have obtained thus far.

- Review again the responses from all the individuals you have interviewed in the industry. Make sure that you have not overlooked any significant factors. A comment that a distributor made earlier in the process may not have meant much to you at the time, but now, after you have communicated with upper-level executives in the industry, the distributor's comment may mean something quite different to you.
- Make sure that you have thoroughly investigated the background of the company you've chosen, and its most recent practices. It is easy to fall into the trap of having a great reception from a company, everyone and everything is positive, but then one to two years later you discover that you have made a terrible mistake. The company has failed to do what it promised, and you have lost other market opportunities in the meantime. Try to avoid this situation by

checking out the company's reputation in the industry, regardless of how good a feeling you have about the company. Contact at least one other person who has done business with the company from the outside and talk to them about their experience.

- Decide now whether you are going to continue this negotiation process solely on your own, or if you think there is even a slight chance of involving professionals, such as an inventor's agent or patent attorney who specializes in license negotiation. If you have any doubts about going through this process alone, now is the time to utilize these professionals. Even if you don't end up eventually hiring someone to do the negotiation, you may want to purchase consultation time and bring him or her up to speed about your project. In this way, if you get stuck and need to enlist a professional, you are not bringing the person in cold; rather, he or she is already primed and ready to go to bat for you. (I realize that many inventors don't want to give up a percentage of their proceeds after having gone this far through a project; however, 90 percent of something is better than 100 percent of nothing. Any good, experienced negotiator should be able to enhance a deal and be worth the 5 to 20 percent that his or her involvement may cost you. Attorneys and some negotiators will perform this function on an hourly basis, as well.)

Selecting the Final Candidates

I like to start negotiation with the top candidate first. The top candidate would be the one who you believe offers the greatest distribution and market potential for your invention, has the appropriate resources you need to complete the project, and is willing to consider offering appropriate payment for your invention. In addition, this company should have a good reputation in the trade (with no serious marks against it), have a champion within the company with whom you feel comfortable doing business, and be a company with whom your trusted advisors feel comfortable as well.

In the event you have a clear candidate, you can proceed with the exercise in this chapter to establish the specific terms of a potential arrangement or contract.

However, if you are still unsure about which candidate is best, refer to Review of Final Candidates in Appendix D. There you will do a cross comparison of your list of final candidates as found in column E on the M/LSW in Appendix C. You may find it helpful to have all the appendix comparison charts in front of you as you enter this final deliberation. These include: Distribution Chart, Appendix B; M/LSW, Appendix C; and the Review of Final Candidates, Appendix D.

If you have done all the exercises in this workbook, the appropriate strategy for effectively commercializing your invention should begin to unfold. It is my experience that at this stage, one or two companies will rise to the forefront as obvious candidates.

Don't force decisions at this point. If you are still uncertain, or feel uncomfortable about a prospective company, this is the time to reevaluate, and gather additional information if necessary. This is also an excellent point in time to consult with one or more professionals who have experience with these types of negotiations.

FACE-TO-FACE NEGOTIATION

Professional Assistance

Having someone else negotiate for you can be especially valuable if your negotiations occur face-to-face. Picture this: You are in a meeting, tensions high, blood pressure up; the company is throwing terms at you to consider, and not only do you need to respond to these comments, you also have to think about repercussions, alternatives, and so on. As you go through potential contract terms, item by item, you don't always have a night to sleep on it. If the company is ready to sign a deal, it is nice to get it done as soon as possible, before something comes up to nix the deal.

If you have a good professional negotiator, he or she will be the one communicating with the company while you sit back in the chair listening and thinking. You can think more clearly this way, take notes, and have pauses in the meeting to discuss the terms with your negotiator, even in private. I have found often that it is better for the inventor to use a negotiator; this way you can avoid the mistake of letting your emotions get in the way or hastily making decisions in the heat of the action.

Use of Attorneys: A Warning

All too often, I have witnessed an inventor's attorney make a deal so complicated that the company's key decision maker chooses to pass on the project. Remember, you make the final decision, not your attorney! You will be the big loser if a deal is crushed. Stand firm and don't be afraid to seek the advice of a professional licensing agent for a second opinion if you feel that your attorney is being overprotective.

For most invention projects, I try to avoid using an attorney for negotiation too prematurely. For one reason, if you bring in an attorney at the onset of negotiation, the company will feel compelled to use its attorney as well. Now the project is suddenly out of the hands of the key decision maker, who is your champion, and in the hands of the company's attorney, who may be completely indifferent as to whether your project succeeds or not. If the basic contract terms such as royalty rate and minimum performance were not already established, however, there is extra opportunity for the attorney to advise against certain of these aspects of a potential deal between the parties. Avoid such a scenario.

I normally find it advantageous instead to arrive at an agreement on the basic terms with the key decision maker in the company. They can then present these terms to an attorney for the purpose of simply buttoning up the final legalese on the contract.

During this process I typically get advice from an attorney who is working for me in the shadows, and as an agent I do the negotiation with the company. Then when the company gets its attorney involved, I bring my attorney into the open.

Selecting a Negotiator

In one respect, I tell you how beneficial it is to use a negotiator, even early in this process. On the other hand, I warn you about using an attorney for negotiation too early in the process. So, does this mean you should not use an attorney to help you with negotiation? Not necessarily.

In my experience it is hard to find good proficient negotiators, and especially when you limit the field to patent attorneys. Therefore you must put in the extra effort to prequalify any person whom you will trust with negotiation. The success of your entire project may rest with your negotiator.

In order to prequalify a negotiator, I believe the most important factor is their track record. What kind of experience does this person have with negotiating business contracts and licensing agreements? There is little substitute for experience, and as such I prefer a negotiator with a significant reference list of successful contracts for which they were the lead negotiator.

Prior to Negotiation

Let's assume that you have a clear candidate with whom you want to do business. This should be the company that you feel will do the best job at commercializing your invention. Even if your top candidate has expressed a keen interest in your invention, the company may want to wait before entering into a contract. The company may want to test-market the invention, produce working models for display at a trade show, send samples to sales representatives to "tickle" the marketplace, and/or present your invention to the purchasing agents of its major accounts. In this event, you may choose to negotiate a simple license option, or a first right of refusal, to give the company time to complete its work.

In other words, you would give the company time to decide if they want to make a formal offer for your invention. In consideration for giving them time to complete their tests, you may charge the company a fee. Such an option fee will vary depending on the value of your invention, the time duration of the option, and the company's willingness to pay. Offering an option is a very common business practice. It gives the company a chance to determine what it's getting into before making a major commitment, thereby lowering their financial risk.

If the company is reluctant to enter into a formal agreement at this point, at minimum, you should try to obtain a letter of intent from the company and place a time limit on its decision-making activities. Unless you have a hallmark invention worth millions, a letter of intent may be the best way to go. It all depends on what's at stake financially. Inventions of greater value generally require more formality when dealing with contracts.

For example, if you have a simple invention, you may want to forgo many of the points listed ahead and go straight to a simple license agreement like the one outlined on page 149 of *The Inventor's Bible*, or some variation thereof. Remember not to lose sight of the forest for the trees. I have seen many a deal fail because the inventor wanted to nitpick fine details about agreements and exceeded the company executive's threshold of patience.

Sometimes key executives have an attitude that they are doing you an extreme favor to license your invention. They may realize that their company is in the absolute best position to proceed in the market. Avoid becoming defensive at the expense of your deal. If this is the best company, and no one else comes close, you certainly want to do more listening than talking, and more agreeing than disagreeing.

Hint: It is not uncommon for company executives to handle orders for products in the magnitude of millions of dollars, based on a handshake with their suppliers or with their major sales outlet. When you come along with a product or business arrangement worth a quarter of what they are used to dealing with, you will only be able to maintain their level of patience for so long in negotiating the finite details of a long, drawn-out contract before the company executive decides to move on to a different project, leaving you in the dust. Again, don't let this happen to you.

If you notice a personality or ego conflict with the company executive, by all means immediately turn this negotiation over to a qualified professional.

Contract Terms

The process of establishing the contract terms will normally happen in one of two different ways. For one, you may sit down in person or over the telephone discuss each term, item by item, with your champion in the company and come to a meeting of the minds. Alternately, the champion may want you to jot down your preferences for each of these terms, send them to him or her, and wait for a counterproposal. You would continue this process until you reach a consensus.

Another effective method is to propose in writing your preferences on some of the terms and leave spaces for the company key executive to fill in the blanks on other terms. Here are terms you will need to consider.

Certain terms may not apply to your circumstance. Choose the terms from the list below that are most crucial to your deal.

- The invention: Will you include future improvements or modifications? Which patents? Which product variations?

____

____

- Territory: What is the geographic territory under this license?

____

- Field of use: Are there any limitations regarding markets; make, use, or sell; and so on?

____

____

- Grant: Will it be exclusive or nonexclusive? Are assignments or sublicenses foreseeable?

____

- Royalties: What will be the royalty rate and any other reimbursement paid to the inventor?

____

____

- Time of payments: When will you receive royalties?

____

- Initial fee/advance: When will you receive lump-sum payments? How much?

____

____

- Performance: What are the minimum performance standards, ramp-up timetables, and associated penalties?

____

____

____

- Patent fees: Will the company pay for your patent and/or patent maintenance fees?

  ____

- Reports: When will you receive written reports, and what details will be included?

  ____

- Duration: What is the life of this contract?

  ____

- Packaging: List special requirements here, if appropriate; use of copyright.

  ____

- Marking: Is it necessary to mark the product and/or the package with your patent number, trademark and/or copyright symbol? Use your attorney's advice with regard to marking requirements.

  ____

- Specifications: Are there special manufacturing tolerances or other special requirements that the manufacturer must follow when producing your invention?

  ____

- Quality control: Are there any special requirements?

  ____

- Inspection: How often might you inspect the manufacturer's production facility, and how often might you inspect the company's financial records to audit royalties due you?

  ____

- Trade secrets: How will you define the working relationship of information sharing between you and the company? (You may slip in a confidentiality clause here.)

  ____

- Enforcement: Who will be responsible for enforcing patent rights; how will expenses and revenues from any litigation be shared between the parties?

____

- Non-compete: Do you need to sell existing inventory and how will this affect the ongoing arrangement with the company?

____

- Termination: Once triggered based on breech of other terms, how will you proceed?

____

____

- Samples: How many will the company send you, how often, and at whose expense?

____

- Special terms: List special terms specific to your project here.

____

____

____

____

____

After you have concluded a number of rounds of communication with the company necessary to address basic terms such as royalty rate, initial payment, minimum performance, duration of contract, and other terms that are important to you, highlight or circle those terms that require additional negotiation. Pay attention to those areas where you are more willing to compromise, and be prepared to give in to some demands in favor of winning the terms of greater importance to you.

You may find yourself having successfully gotten all the way to this point in negotiation only to be in nonnegotiable disagreement on one very basic term that will make or break the deal. I find that the minimum performance clause in particular is a great deal killer. It is not uncommon for a company to want to bind you with exclusive rights for the twenty-year life term of your patent and yet offer miserably low minimum performance standards. In this event, if the company does not perform well enough, you are potentially stuck for the life of the patent.

If this happens with any critical and nonnegotiable term of the proposed agreement, you may have no other alternative than to scrap negotiations with this company and go on to the next best qualified licensing candidate. Then you will be going through this entire negotiation process again.

This underscores the importance of getting a rough idea as to the company's vision of the market potential for your invention well before you enter the negotiation stage. All too often inventors get so excited about receiving a positive response from a company that they forget to ask some very basic questions, and such an omission can come back to haunt you. The last thing you want is to have incurred an extra $3,000 to $5,000 of expenses for your licensing agent or attorney to help you through this process, only to have it nixed because you didn't initially gather the appropriate market information.

Attorney's Final Review

Let's assume that you have reached the point where you and the company's key decision maker have agreed to all the terms most important to both of you. You are now in an excellent position to turn this information over to your patent attorney, the one who specializes in handling the negotiation of license contracts. The attorney can complete the more detailed legal terms. Be sure to have a discussion with your attorney about his or her recommendations. You will normally not know whether you have a meeting of the minds on these specific terms until you have gotten a reply from the other party's legal counsel.

It is not uncommon at this stage for your attorney to list the terms in which you are most interested and receive few changes or modifications from the other side. Small- to medium-size companies, in particular, often attempt to minimize their attorney expenses, and therefore direct their attorneys to make no more modifications than necessary to establish a clean deal. Again, this underlines the importance of getting your key decision maker to agree to as many of the terms of this contract as possible beforehand.

Sometimes key decision makers will sign these agreements without a review by their legal counsel. They have seen so many of these common terms and clauses before that they don't find such consultations are necessary. In this event, you are off and running!

WORKING WITH LARGE COMPANIES: A DISCLAIMER

If you are dealing with a multinational corporation, all this discussion about establishing terms with key decision makers may be moot. The fact is, you will be dealing with the company's corporate legal department from the beginning, starting with the signing of a disclosure agreement prior to submitting your invention. Many multinational companies have this corporate legal shield to help reduce their legal liability. However, even with a multinational company or corporation, you may still be able to negotiate the basic terms with a key decision maker. He or she will then most certainly turn this information over to the legal department for final approval of any contract.

A Final Word

I wish you all the luck in getting your invention out into the marketplace. Although this workbook serves as a guideline to assist you, the real effort comes from you. Your hard work and toil will most certainly bring you to some form of resolution. The end result will either be the negotiation of a successful deal, or you will know some very good reasons why the commercialization of your invention was not in the cards.

Let's say that you have negotiated successfully and established satisfactory contract terms including cash up front. The battle is still not over. Your product or technology must successfully compete. Everything will have to go just right with manufacturing and distribution, and hopefully the markets won't take an adverse dive against you.

Either you or your agent will be well served to make contact with your licensee once per calendar quarter. This should not be to merely check on royalties, rather you should keep abreast of activity involving the manufacturing and marketing of your invention. Such contact helps you to adapt to inevitable changes in the conditions that will affect your project over time.

If you negotiate a great deal, and your invention is still on the market in three years, you have the proverbial pat on the back from me because you will have gone far beyond hundreds who have tried.

Good luck to you.

Invitation

If you successfully negotiate a deal, having used the exercises in this workbook to help you along, please notify me, preferably by U.S. mail, or by email. I would like to compile a list of those successful inventors who make this system work, and I may produce a story in the future about your heroic efforts.

APPENDIX A

BOOKMARK

See pages 66–74 in *The Inventor's Bible* for more information about how to utilize these valuable resources.

Trade Show Information

I cannot overemphasize the value of attending your industry's major national trade show or conference.

Trade Show Listing Chart

Use this worksheet to summarize information about upcoming trade shows. Compile what you learned from your interviewing, and from any other sources.

Trade show: ______________________________

Location: ____________________ Dates: ______________

Entry requirements to attend: ______________________

Number of exhibitors: ________ Trade served: ______________

Phone number: ______________________________

Website: ______________________________

Workshops/conferences offered, date, and cost: ______________

____

____

Trade Show Interviewing

At major national trade shows, you may have an average of twenty to forty seconds to spend at each of the two to three thousand exhibits. Therefore, you must quickly ascertain whether the exhibitor is a candidate with whom to do business. First, determine if your invention is compatible with the company's product line. If it is, enter the exhibit and ask to speak with whoever considers new products in your product category. Then ask the questions below. (**NOTE:** Try to memorize these questions, keep the answers in your mind, and only take essential notes, such as names and phone numbers, during the interview. The bulk of the interview requires eye contact. Fill in the blanks later.) Make several copies of the Trade Show Worksheet that follows. In lieu of copying contact information, staple your interviewee's business card to the worksheet.

A. Is the company interested in new product ideas relating to (your area of invention or invention catagory)?

B. Does the company have any history of working with outside inventors and paying them royalties? (Note: If the answer to either of these first two questions is no, kindly ask what other companies they recommend. If both answers are yes, continue with the following questions for as long as their patience allows.)

C. Does the person with whom you are speaking make the final decision regarding licensing inventions from the outside? If not, who does?

D. Should you be dealing with the final decision maker directly? (Note: Many times the vice president of marketing or general manager for a company may be the person who spearheads the introduction of new ideas into the company and then, after their approval, they pass them on to the owner or other people who make the final decision. In other words, the vice president makes the recommendation; the president makes the decision. If this is the company's style and if you try to circumvent it by going directly to the owner, you may sabotage your ability to license your invention. So essentially, you need to learn the ropes, and asking directly is not a bad way to determine who influences whom.)

E. What are examples of new products the company has introduced in the marketplace that were acquired from outside inventors?

F. What kind of success has the company had recently with new product introductions?

G. What is the company's experience in dealing with outside inventors? Does the company prefer to pay licensing fees or purchase the patent outright?

H. Is there an example of royalty rates for this product category? Does the company ever pay in advance to help offset an inventor's expenses?

I. Would the company be interested in looking at your invention?

J. Whom should the invention be initially sent to? (You will need contact information.)

K. What kind of information would the company want to see (photos only, working model, samples, and so on)?

L. How much time does the company generally need to make a decision to go further?

In the event you do choose to reveal your invention to a company representative, you can get specific feedback using some of the questions suggested in Chapter 3 in "Getting Specific Feedback about Your Invention." Do not carry anything larger than a breadbox into the trade show. Your invention should be totally concealed in a case so as not to appear like you are trying to sell something. Beyond this, use photos or illustrations. You can alternatively meet the company representative outside the show, after hours, if they are excited about seeing your invention.

The bottom line here is that the response usually goes in one of two different directions. Either you get a positive response and start a dialogue about how to get to the next step with the company, or you get a negative response.

If at any point during the interview the response turns negative, ask for a referral to another company that they think may be in a better position to commercialize your invention. This is a good time to get a referral to a key decision maker in another company since many people know each other in any specific industry. Also, be sure to ask if there are any aspects to your invention that they believe need to be changed or improved. This may be critical. As you interview companies in this manner, the process may yield valuable feedback on ways to enhance your invention so that you can refine it for presentation to the company you ultimately choose to approach.

Trade Show Worksheet

Trade show: ______________________ Date: ______________________

(Attach your interviewee's business card here.)

Company: ______________________ Interviewee: ______________________

Brand name: ______________________ Contact person: ______________________

Contact phone: ______________________ Contact title: ______________________

Email: ____

URL: ____

History with inventors: ____

____

Interests/comments: ____

____

APPENDIX B

DISTRIBUTION CHART

This distribution chart is where you will summarize certain information discovered in your first level of market research. Various action steps in the workbook direct you to record answers to select questions on this chart.

NOTE: If there are hundreds of small independent stores or sales outlets for your invention, lump these together as "independent" and list any prominent manufacturers who serve these independents.

There is a place on the chart to make note of "invention configuration." You may find that different versions of your invention will be appropriate for different types of sales outlets. You may have an all-metal heavy-duty version of your invention appropriate for industrial or military uses. The lightweight, inexpensive plastic version of your invention may be more appropriate for mass merchandisers and other retail outlets. Invention configuration is where you would distinguish which version of your invention is appropriate for that particular sales outlet.

	FIRST OUTLET	**SECOND OUTLET**	**THIRD OUTLET**
Retail outlet name	______	______	______
Corporate name	______	______	______
Number of outlets	______	______	______
Territory served	______	______	______
Invention configuration	______	______	______
Distributors/wholesalers	______	______	______
(for this outlet)	______	______	______
	______	______	______
Manufacturers	______	______	______
(serving this outlet)	______	______	______
	______	______	______
	______	______	______
	______	______	______
	______	______	______

APPENDIX C

Manufacturer/Licensee Selection Worksheet

How to Use This Worksheet

In **Column A,** list those companies that have the best market presence. These will be found during your First-Level market research, and from the Distribution Chart in Appendix B.

In **Column B,** list the companies most referred by your interviewees during all levels of your market research. These company names will reoccur throughout the interviewing process.

In **Column C,** summarize the results from Columns A and B by listing the best manufacturer/licensee candidates. You will learn which are the best candidates from your interviews in the first and second levels, and from attending trade shows or conferences. This list will be your targets to communicate with in the third level of market research.

In **Column D,** after you receive responses from companies in your third level of market research, you will further narrow down the list of potential manufacturers/licensees. List these semifinalists here. Although the companies in this column expressed an interest in your invention category, more screening is necessary.

In **Column E,** after your manufacturer/licensee candidates have reviewed the essence of your invention, your choices will become more clear, thus leading to this list of finalists. These are the companies with which you will strongly consider negotiating a final arrangement.

In **Column F,** as this process evolves, new company names may come into focus, while others will drop off—add here new candidates and those companies for which you may want to investigate secondary markets.

In **Columns E and F,** the final companies listed in these two columns can be transferred to the Review of Final Candidates worksheet in Appendix D. The review process will help to determine the order in which you choose to proceed.

Secondary market is where you record distinctly different markets, and/or markets for different applications or configurations of your invention.

MANUFACTURER/LICENSEE SELECTION WORKSHEET

A. MARKET PRESENCE

SECONDARY MARKET

B. MOST REFERRED

SECONDARY MARKET

C. BEST CANDIDATES (INVENTOR FRIENDLY)

SECONDARY MARKET

D. SEMIFINALISTS

SECONDARY MARKET

E. FINALISTS

SECONDARY MARKET

F. OTHER MANUFACTURERS

SECONDARY MARKET

APPENDIX D

REVIEW OF FINAL CANDIDATES

Use this worksheet to review your final candidates. Try to fill in as many blanks as possible before going into any formal negotiation or before offering your invention for sale. If any of the following important terms are not yet defined, recontact your champion in the company and get the answers. Discussion about potential contract terms are best guesses at this point and not necessarily a firm commitment by the company.

	FIRST CHOICE	SECOND CHOICE
A. Company name:	____________	____________
B. Geographic sales territory:	____________ ____________	____________ ____________
C. Applications: (Which variations of your invention will the company handle?)	____________ ____________	____________ ____________
D. Product line extension: (Which of your improvements is the company interested in?)	____________ ____________	____________ ____________
E. Markets:	____________ ____________ ____________ ____________	____________ ____________ ____________ ____________

	FIRST CHOICE	SECOND CHOICE
F. Major stores/ distribution outlets:		
G. Exclusivity: (Does this company want an exclusive arrangement?)		
H. Sales projection (range):		
I. Minimum sales projection:		
J. Advanced payment (possible range):		
K. Royalty rate (possible range):		
L. Timetable for market introduction:		
M. Duration of contract (in years):		
N. Projected retail sale price (for each product variation):		
O. Initial production run (number of units):		
P. Specific resource allocation, if appropriate: (For example, they will invest $50,000 for tooling and $20,000 for packaging design.)		

	FIRST CHOICE	SECOND CHOICE
Q. Resource gap: (List any additional resources required, such as engineering support, marketing support, manufacturing, and so on.)		
R. Positive references:		
S. Other important factors specific to your project:		
T. Who will do the final negotiations? (If you are using a professional, include name, company, and phone number.)		
U. What is the amount of time needed by the parties to decide on this project?		

POCKET GUIDE:

THE LUCKY SEVENTY-SEVEN QUESTIONS FOR SUCCESS

THIS QUICK REFERENCE GUIDE is meant to help those schooled in the teaching of *The Inventor's Bible* to nail down the critical information needed for your invention project. This guide will also serve as a checks-and-balances review for the research you've performed in the workbook. At a minimum, you should be able to answer the following questions about your best manufacturing/licensing candidates.

First Level

1. Summarize: Which manufacturers/brands most closely match the market position suitable for your invention?
2. What drives sales? (Example: mostly price, time saving, color, special features—which ones?)
3. What price ranges/features are most popular?
4. Who supplies them? (Example: their own warehouse, independent distributor, direct from China, and so on.)
5. Catalogs and websites? Identify collateral sales in this category.
6. Trade associations and publications that serve this industry (name, address, URL, telephone)?
7. Trade shows that serve this industry (dates and locations of shows)?
8. Ask for referral to other locations to visit that have similar products/technologies in this category.
9. Ask for referral to another employee or expert who is particularly knowledgeable about this product category. Are they available now?

Second Level

10. If this were their invention, which manufacturers would they prefer to work with and why?

11. Which manufacturers have a better presence in foreign markets; which countries?
12. Which foreign manufacturers are making an aggressive impact in the United States; which sales outlets?
13. What trends do they see for products in this category?
14. What barriers do they see in marketing a product in this category?
15. What features and benefits are driving sales in this product category?
16. Which other experts in this field do they recommend that you speak with? (Example: authors, engineers, professors, consultants, government agencies, other salespeople, and so on.)
17. What are the names, titles, and contact information for any individual employees of the manufacturer that they recommend you contact directly?
18. What are the prominent mail order catalogs for this product category?
19. What are the URL's of the prominent websites offering online sales in this product category?

Getting Feedback about Your Invention

20. Have they ever seen anything like your invention? What did they see, where, and when?
21. What do they like most about your invention?
22. Which manufacturer would be most likely to want this product in their product line?

Questions at Trade Shows

23. Is the company interested in new product ideas relating to (your area of invention or invention catagory)?
24. Does the company have any history with working with outside inventors and paying them royalties?
25. Does the person with whom you are speaking make the final decision regarding licensing inventions from the outside? If not, which other people do?

Third Level and Industry Experts

26. What do they think about the invention concept?
27. Would an invention like this support your company's market position, and if so, where would an invention like this fit into your product line? (In which product category, with which other items, in which division of the company, and so on?)

28. What drives sales in this product category? (Example: price, packaging, features—which ones?)
29. What are the ideal sales outlets? (Example: mass merchandisers, specialty stores, mail order, and so on.)
30. How strong is the company in these sales outlets?
31. What is the general retail price range for a product like this? (At least get a ballpark figure: $5 to $10, under $1, $100 to $150, and so on.)
32. What is the potential sales volume for a product like this? (Example: 5 to 10 thousand annually, 10 to 50 million annually, and so forth—just get a very broad ballpark here so at least you know how many zeros are in the person's mind.)
33. Where do they see the sales? (Example: nationally, regionally, internationally; which countries, and so forth.)
34. What might be the product life cycle? (Example: five years max in its current form, ten years plus, indefinite, and so on.)
35. What would be the time frame for introducing your invention, that is how long would it take to get it to the broader market and how long would it take to ramp up sales? (Example: 5 thousand the first year, 20 thousand the second year, 100 thousand the third year, and so on.)
36. What experience do they have with working with outside inventors, that is, how many outside inventors have they worked with in the past, and when? (This is crucial information.)
37. Has the company ever paid inventors for unpatented inventions?
38. What royalty rates are typical in this industry and/or for this company? (Example: 5 percent of net sales, 2 to 3 percent, paid by number of units only, and so forth.)
39. Does the company typically prefer to purchase patents outright, and do they ever make advance payments?
40. What experience do they have with manufacturing products similar to yours? Where are such products manufactured?
41. What experience and resources do they have for engineering, quality control, and service after the sale, and where do they see their strong suits?
42. How finished do the drawings or working models need to be? (Example: Are they looking for a "looks like" model, or a "works like" model, or both?)
43. Do they expect to make design improvements and refinement in-house, expect this solely from the inventor, or prefer collaboration between the two?
44. In what general areas are the company's strong suits? (Example: marketing, manufacturing, engineering, and so on.)
45. What market position has the company built for itself? (Example: superior value, low price, highest quality, superior service, and so on.)

46. How long has the company been in business, and how long under current management? Are there any anticipated changes in management or ownership?
47. What departments or what people within the company need to be influenced to make the final decision about accepting your invention? Is there any one area that needs convincing more than another?
48. What other companies do they think would be a better contact for you, that is in a better position to help you, or in a better market position?
49. What are the greatest barriers or resistance to introducing a new invention like yours, or specifically, what problems do they see with your invention?
50. What companies offer the stiffest competition in this product category? What are their names and in what cities are they located?

Questions for Recontact

51. What is the overall impression of the invention?
52. Was the information you sent them clear and did it thoroughly explain the advantages and benefits of your invention?
53. What additional information may they need to better fully understand the invention?
54. Are they interested in further considering your invention?

For a Negative Response

55. Have they ever seen anything like your invention and what is the closest thing to it they have seen?
56. Does the invention fit the company's market or support its market position?
57. Do they see problems with a product like this and what are they?
58. Do they see barriers in the market for introducing the invention?
59. Do they see possible modifications that would make the invention more acceptable? (Example: different size, different features, and so forth.)
60. Do they see applications for your invention in other markets? Which ones?
61. Names of other companies that may be in a better position to commercialize the invention? (Get city and state if possible.)

For a Positive Response

62. What is it about the features and benefits that is most appealing? (Get into specifics here.)
63. Do they need additional information to enhance their further investigation, and what do they need?

64. Would they like to meet you in person and see the invention?
65. What is the internal procedure for further review? Who specifically will be doing the review?
66. Who reports to whom? Who makes the final decision?
67. When will the review be complete?
68. Is it okay to contact the people doing the review to provide further clarification?
69. What kind of sales volume do they envision for this product over time?
70. Are they more interested in an outright purchase, or in paying ongoing royalties?
71. Are they interested in an exclusive arrangement? In which markets, for which applications?
72. For how long are they interested in an exclusive? (Example: three years, ten years, life of patent, and so forth.)
73. What foreign markets do they foresee for your invention and in which of these markets does the company have a stronghold or good distribution?
74. What will be the deciding factor(s) that will determine whether they want to proceed with this project? (Example: If the sales department embraces it, they will go for it; engineering must approve it, and then it goes straight to the owner of the company for final consideration; and so on.)
75. What is the timeline for learning the final determination?
76. Is there anything that you, the inventor, can do in the meantime to support or positively influence the final decision? (Example: They may suggest more test data of a particular nature, results of the test market, a working model presented in person, testimonials, and so on.)
77. What is their gut feeling about the chances of this project proceeding?

INDEX

J

K

L

U

V

W

Y

ABOUT THE AUTHOR

Ronald Louis Docie, Sr., started peddling his family's garden vegetables from door to door at age five. Today, he teaches the art of sales and marketing under the auspices of Docie Marketing, a division of Hopewell Cooperative, Inc., a company dedicated to educating inventors and providing commercialization services. While acquiring funding for and licensing his first invention (an automotive accessory that still sells throughout the world in stores like Kmart and Wal-Mart) at age twenty-four, Docie handled marketing projects for entrepreneurs who had invention-based businesses.

Over the past thirty-three years, Docie has negotiated over fifty licenses and contracts for inventions and invention-based projects. He is a speaker and workshop leader at inventor's conferences, and his articles are regularly published in trade journals and on the Internet. A three-term president of the nonprofit Ohio Inventors Association, creator of the Ohio Inventors Contest, co-developer of the Ohio Inventors Resource Guide (jointly with the State of Ohio, Department of Development), Docie has testified on behalf of independent inventors in congressional subcommittees in Washington, D.C., and was a participant in the National Congress of Inventor Organizations and the original United Inventors Association (UIA). Most recently, Docie created and developed DIMWIT.com, a website for inventors.

Photo © 2001 by Chris Eaton

An avid auto racing driver and pilot, Docie's latest project is the restoration of a log cabin using traditional, pioneer-era tools, including a working mule team. Docie lives in Athens, Ohio.